Leadership in Energy and Environmental Design

LEED® O&M Practice Exam

Operations & Maintenance

David M. Hubka, DE, LEED AP

Professional Publications, Inc. • Belmont, California

How to Locate and Report Errata for This Book

At PPI, we do our best to bring you error-free books. But when errors do occur, we want to make sure you can view corrections and report any potential errors you find, so the errors cause as little confusion as possible.

A current list of known errata and other updates for this book is available on the PPI website at **www.ppi2pass.com/errata**. We update the errata page as often as necessary, so check in regularly. You will also find instructions for submitting suspected errata. We are grateful to every reader who takes the time to help us improve the quality of our books by pointing out an error.

LEED® and USGBC® are registered trademarks of the U.S. Green Building Council. PPI® is not affiliated with the U.S. Green Building Council (USGBC) or the Green Building Certification Institute (GBCI), and does not administer the LEED AP program or LEED Green Building Rating System. PPI does not claim any endorsement or recommendation of its products or services by USGBC or GBCI.

Energy Star® (ENERGY STAR®) is a registered trademark of the U.S. Environmental Protection Agency (EPA).

Also available for the LEED exams at **www.ppi2pass.com/LEED**:
LEED Prep O&M: What You Really Need to Know to Pass the LEED AP Operations & Maintenance Exam
LEED O&M Flashcards: Operations & Maintenance
LEED GA Practice Exams: Green Associate
LEED Prep GA: What You Really Need to Know to Pass the LEED Green Associate Exam
LEED GA Flashcards: Green Associate

LEED O&M PRACTICE EXAM: OPERATIONS & MAINTENANCE

Current printing of this edition: 1

Printing History

edition number	printing number	update
1	1	New book.

Copyright © 2009 by Professional Publications, Inc. (PPI). All rights reserved. No part of this publication may be reproduced, stored in a retrieval system, or transmitted, in any form or by any means, electronic, mechanical, photocopying, recording, or otherwise, without the prior written permission of the publisher.

Printed in the United States of America

PPI
1250 Fifth Avenue, Belmont, CA 94002
(650) 593-9119
www.ppi2pass.com

ISBN: 978-1-59126-181-0

Table of Contents

Preface and Acknowledgments v

Introduction xi
    About the LEED Credentialing Program xi
    About the LEED AP Operations & Maintenance Exam xi
    Taking the LEED Credentialing Exams xiii
    How to Use This Book xiii

References vii
    Primary References for Part One of the LEED AP Operations & Maintenance Exam vii
    Secondary References for Part One of the LEED AP Operations & Maintenance Exam . . . viii
    Primary References for Part Two of the LEED AP Operations & Maintenance Exam viii
    Secondary References for Part Two of the LEED AP Operations & Maintenance Exam ix

Practice Exam Part One 1

Practice Exam Part Two 23

Practice Exam Part One Solutions 47

Practice Exam Part Two Solutions 67

Preface and Acknowledgments

This book, which is written to help you study for and pass the LEED AP Operations & Maintenance credentialing exam, is a product of my passion for green building. Soon after I began designing mechanical systems, I recognized that many contractors, engineers, product vendors, and architects desired a fundamental understanding of LEED. As a result, I began creating hour-long seminars discussing the impact of buildings on the environment and the growing presence of LEED for the benefit of local building professionals. The success of these local seminars paved the way for me to present LEED exam prep training seminars throughout the country. The problems in this book, which are unique, based on my experiences, and representative of those on the actual exam, are outgrowths of those presentations and seminars.

I wish to acknowledge those who have helped me create this book. First, thanks go to the incredibly talented professionals working at Total Mechanical. Their combined experience across all mechanical trades has proven to be an invaluable resource for me; without their assistance I would not have the broad knowledge of LEED that I have today. Thanks also go to Mike Hyde, Total Mechanical's Chief HVAC Engineer. He continually provides forward thinking solutions to our LEED projects. I am deeply appreciative of the Mechanical Service Contractors of America (MSCA) for hiring me to provide my first LEED exam prep training, and of Barb Dolim, MSCA Executive Director, for providing direction as I developed the training. I would also like to thank all of my LEED seminar attendees, who continually contribute to my understanding and appreciation of the rewards and challenges of green building.

Next, thank you to those at PPI who helped in the process of creating this book, including director of new product development Sarah Hubbard, director of production Cathy Schrott, editorial assistant Courtnee Crystal, and typesetter and cover designer Amy Schwertman.

Finally, I would like to thank my wife, Dana, for her support throughout the entire process.

Despite all of the help and support I had, any mistakes you find are mine. If you find any mistakes, please report them through **www.ppi2pass.com/errata**. Corrections will be posted on the PPI website and incorporated into this book when it is reprinted.

Good luck on the exam and in all your green building efforts.

David M. Hubka, DE, LEED AP

Introduction

About the LEED Credentialing Program

The Green Building Certification Institute (GBCI) offers credentialing opportunities to professionals who demonstrate knowledge of Leadership in Energy and Environmental Design (LEED) green building practices. *LEED O&M Practice Exams: Operations & Maintenance* prepares you for the LEED AP Operations & Maintenance exam and part one of each of the LEED Accredited Professional (AP) specialty exams.

GBCI's LEED credentialing program has three tiers. The first tier corresponds to the LEED Green Associate exam. According to the *LEED Green Associate Candidate Handbook*, this exam confirms that you have the knowledge and skills necessary to understand and support green design, construction, and operations. When you pass the LEED Green Associate exam, you will earn the LEED Green Associate credential.

The second tier, which corresponds to the LEED AP specialty exams, confirms your deeper and more specialized knowledge of green building practices. GBCI currently has planned five tracks for the LEED AP exams: LEED AP Homes, LEED AP Operations & Maintenance, LEED AP Building Design & Construction, LEED AP Interior Design & Construction, and LEED AP Neighborhood Development. The LEED AP exams are based on the corresponding LEED reference guide and rating systems and other references. When you pass the LEED Green Associate exam along with any LEED AP specialty exam, you will earn the LEED AP credential.

The third tier, called LEED AP Fellow, will distinguish professionals with an exceptional depth of knowledge, experience, and accomplishments with LEED green building practices. This distinction will be attainable through extensive LEED project experience, not by taking an exam.

For more information about LEED credentialing, visit **www.ppi2pass.com/LEEDhome**.

About the LEED AP Operations & Maintenance Exam

The LEED AP Operations & Maintenance exam (and the practice exam in this book) contains 200 problems. The first part of the exam (whose specifications are identical to those of the LEED Green Associate exam) contains 100 questions that test your knowledge of green building practices and principles, as well as your familiarity with LEED requirements, resources, and processes. Accordingly, GBCI categorizes the exam questions into the following seven subject areas.

- *Synergistic Opportunities and LEED Application Process* (project requirements; costs; green resources; standards that support LEED credit; credit interactions; Credit Interpretation Requests and rulings; components of LEED online and project registration; components of LEED score card; components of letter templates; strategies to achieve credit; project boundary; LEED boundary; property boundary; prerequisites and/or minimum program requirements for LEED certification; preliminary rating; multiple certifications for same building; occupancy requirements; USGBC policies; requirements to earn LEED AP credit)
- *Project Site Factors* (community connectivity: transportation and pedestrian access; zoning requirements; development: heat islands)
- *Water Management* (types and quality of water; water management)
- *Project Systems and Energy Impacts* (environmental concerns; green power)
- *Acquisition, Installation, and Management of Project Materials* (recycled materials; regionally harvested and manufactured materials; construction waste management)
- *Stakeholder Involvement in Innovation* (integrated project team criteria; durability planning and management; innovative and regional design)
- *Project Surroundings and Public Outreach* (codes)

The second part of the exam contains an additional 100 questions that test your knowledge of subject areas unique to the operations and maintenance of a building. These are the subject areas covered in the second part of this book's practice exam. GBCI has identified these as follows.

- *Project Site Factors* (lighting development; green management; climate conditions)
- *Water Management* (water treatment; stormwater; irrigation demand; chemical management)
- *Project Systems and Energy Impacts* (energy performance policies; building components; on-site renewable energy; third-party relationships: requirements and alternate rating systems; energy performance measurement; energy tradeoffs; energy sources; energy usage; specialized equipment power needs)
- *Acquisition, Installation, and Management of Project Materials* (building reuse; rapidly renewable materials for facilities alterations and additions; food materials; materials acquisition; chemical management policy and audit; environmental management plan)
- *Improvements to the Indoor Environment* (minimum ventilation requirements; tobacco smoke control; indoor air quality: ventilation effectiveness, pre-construction, during construction, before occupancy, and during occupancy; low-emitting materials; indoor/outdoor chemical and pollutant control; lighting controls; thermal controls; views; types of building spaces)
- *Stakeholder Involvement in Innovation* (design workshop/charrette; earning credit through innovation; education of a building manager)
- *Project Surroundings and Public Outreach* (infrastructure; zoning requirements; government planning agencies; public-private partnership: incentives and opportunities; traffic studies; reduced parking methods; ADA/universal)

Introduction

Taking the LEED Credentialing Exams

To apply for a LEED credentialing exam, you must agree to the disciplinary policy and credential maintenance requirements and submit to an application audit. To be eligible to take the LEED Green Associate exam, one of the following must be true.

- Your line of work is in a sustainable field.
- You have documented experience supporting a LEED-registered project.
- You have attended an education program that addresses green building principles.

To be eligible to take a LEED AP exam, you must have documented experience with a LEED Registered Project within the three years prior to your application submittal.

The LEED credentialing exams are administered by computer at Prometric test sites. Prometric is a third party testing agency with over 250 testing locations in the United States and hundreds of centers globally. To schedule an exam, you must first apply at www.gbci.org to receive an eligibility ID number. Then, you must go to the Prometric website at www.prometric.com/gbci to schedule and pay for the exam. If you need to reschedule or cancel your exam, you must do so directly through Prometric.

The LEED credentialing exam questions are multiple choice with four or more answer options for each question. If more than one option must be selected to correctly answer a question, the question stem will indicate how many options you must choose. Each 100-question exam lasts two hours, giving you a bit more than one and a half minutes per question. The bulk of the questions are non-numerical. Because calculators are not allowed or provided, only basic math is needed to correctly solve any quantitative questions. No reference materials or other supplies may be brought into the exam room, though a pencil and scratch paper will be provided by the testing center. (References are not provided.) The only thing you need to bring with you on exam day is your identification.

Your testing experience begins with an optional brief tutorial to introduce you to the testing computer's functions. When you've finished the tutorial, questions and answer options are shown on a computer screen, and the computer keeps track of which options you choose. Because points are not deducted for incorrectly answered questions, you should mark an answer to every question. For answers you are unsure of, make your best guess and flag the question for later review. If you decide on a different answer later, you can change it, but if you run out of time before getting to all your flagged questions, you still will have given a response to each one. Be sure to mark the correct number of options for each question. There is no partial credit for incomplete answers (or for selecting only some of the correct options).

If you are taking both the first tier (LEED Green Associate) and the second tier (LEED AP) exams on the same day, at the end of your first session the computer will ask you if you are ready to take the second tier. You can take a short break at this time. The second tier's two hours begins when you click "yes" to indicate that you are ready.

To ensure that all candidates' chances of passing remain constant regardless of the difficulty of the specific questions administered on any given exam, GBCI converts the raw exam score to a scaled score, with the total number of points set at 200 and a minimum passing score of 170. In this way, you are not penalized if the exam taken is more difficult than another exam. Instead, in such a case, fewer questions must be answered correctly to achieve a passing score. Your scaled score (or scores, if you are taking both tiers on the same day) is reported on the screen upon completing the exam. A brief optional exit survey completes the exam experience.

When you pass the LEED Green Associate exam, a LEED Green Associate certificate will be sent to you in the mail. If you take and pass both exams, a LEED AP certificate will be sent to you in the mail. If you take both exams but pass only the LEED AP exam, you will need to reregister, retake, and pass the LEED Green Associate exam before you receive any LEED credential.

How to Use This Book

There are a few ways you can use this book's practice exam. You can do an untimed review of the questions and answers to familiarize yourself with the exam format and content, determine which subjects you are weak in, and use the information as a guide for studying. (Use the materials identified in this book's "References" section as the basis for your exam review.) Or you can use the exam to simulate the exam experience, either before you begin your study (as a pre-test) or when you think you are fully prepared.

To simulate the exam experience, don't look at the questions or answers ahead of time. Put away your study materials and references, set a timer for two hours, and solve as many questions as you can within the time limit. Practice exam-like time management. Fill in the provided bubble sheet with your best guess on every question regardless of your certainty and mark the answers to revisit if time permits. If you finish before the time is up, review your work. If you are unable to complete the exam within the time limit, make a note of where you were after two hours; but, continue on to complete the exam. Keep track of your time to see how much faster you will need to work to finish the actual exam within two hours.

After taking a practice exam, check your answers against the answer key. Consider a problem correctly answered only if you have selected all of the required options (and no others). Calculate the percent correct. Though the actual exam score will be scaled, aim for getting at least 70% (70 questions) of the practice exam's questions correct. The fully explained solutions are a learning tool. In addition to reading the solutions to the questions you answered incorrectly, read the explanations to those you answered correctly. Categorize your incorrect responses by exam subject to help you determine the areas you need to study. Use the references list to guide your preparation. Though this exam reflects the breadth and depth of the content on the actual exam, use your best judgment when determining the subjects you need to review.

References

The LEED O&M Practice Exam: Operations & Maintenance is based on the following references, identified by the Green Building Certification Institute (GBCI) in its *LEED AP Operations & Maintenance Candidate Handbook*. Most of these references are available electronically. You can find links to these references on PPI's website, **www.ppi2pass.com/LEEDreferences**.

Primary References for Part One of the LEED AP Operations & Maintenance Exam

Bernheim, Anthony and William Reed. "Part II: Pre-Design Issues." *Sustainable Building Technical Manual*. Public Technology, Inc. 1996.

Cost of Green Revisited: Reexamining the Feasibility and Cost Impact of Sustainable Design in Light of Increased Market Adoption. Sacramento, CA: Davis Langdon, 2007.

Introduction and Glossary. *Existing Buildings: Operations & Maintenance Reference Guide*. Washington, DC: U.S. Green Building Council, 2008.

Guidance on Innovation & Design (ID) Credits. Announcement. Washington, DC: U.S. Green Building Council, 2004.

Guidelines for CIR Customers. Announcement. Washington, DC: U.S. Green Building Council, 2007.

LEED for Homes Rating System. Washington, DC: U.S. Green Building Council, 2008.

LEED Technical and Scientific Advisory Committee. *The Treatment by LEED of the Environmental Impact of HVAC Refrigerants*. Washington, DC: U.S. Green Building Council, 2004.

Secondary References for Part One of the LEED AP Operations & Maintenance Exam

AIA Integrated Project Delivery: A Guide. American Institute of Architects, 2007.

Americans with Disabilities Act: Standards for Accessible Design. 28 CFR Part 36. Washington, DC: Code of Federal Regulations, 1994.

"Codes and Standards." Washington, DC: International Code Council, 2009.

"Construction and Building Inspectors." *Occupational Outlook Handbook*. Washington, DC: Bureau of Labor Statistics, 2009.

Frankel, Mark and Cathy Turner. *Energy Performance of LEED for New Construction Buildings: Final Report*. Vancouver, WA: New Buildings Institute and U.S. Green Building Council, 2008.

GSA 2003 Facilities Standards. Washington, DC: General Services Administration, 2003.

Guide to Purchasing Green Power: Renewable Electricity, Renewable Energy Certifications, and On-Site Renewable Generation. Washington, DC: Environmental Protection Agency, 2004.

Kareis, Brian. *Review of ANSI/ASHRAE Standard 62.1-2004: Ventilation for Acceptable Indoor Air Quality*. Greensboro, NC: Workplace Group, 2007.

Lee, Kun-Mo and Haruo Uehara. *Best Practices of ISO - 14021: Self-Declared Environmental Claims*. Suwon, Korea: Ajou University, 2003.

LEED Steering Committee. *Foundations of the Leadership in Energy and Environmental Design Rating System: A Tool for Market Transformation*. Washington, DC: U.S. Green Building Council, 2006.

Primary References for Part Two of the LEED AP Operations & Maintenance Exam

Bernheim, Anthony and William Reed. "Part II: Pre-Design Issues." *Sustainable Building Technical Manual*. Public Technology, Inc. 1996.

Guidance on Innovation & Design ID Credits. Announcement. Washington, DC: United States Green Building Council, 2004.

LEED Online Sample Credit Templates. Washington, DC: United States Green Building Council, 2009.

LEED Reference Guide for Green Building Operations and Maintenance. Washington, DC: United States Green Building Council, 2008.

Secondary References for Part Two of the LEED AP Operations & Maintenance Exam

Architectural Barriers Act. Washington, DC: United States Access Board, 1968.

Energy Star Buildings and Plants. Washington, DC: Environmental Protection Agency, 2009.

Energy Star Portfolio Manager. Washington, DC: Environmental Protection Agency, 2009.

References

IESNA Technical Memorandum on Light Emitting Diode (LED) Sources and Systems. New York: Illuminating Engineering Society of North America, 2005.

Sub-Metering Energy Use in Colleges and Universities: Incentives and Challenges. Washington, DC: Environmental Protection Agency, 2002.

Practice Exam Part One

1. Ⓐ Ⓑ Ⓒ Ⓓ
2. Ⓐ Ⓑ Ⓒ Ⓓ Ⓔ
3. Ⓐ Ⓑ Ⓒ Ⓓ
4. Ⓐ Ⓑ Ⓒ Ⓓ Ⓔ
5. Ⓐ Ⓑ Ⓒ Ⓓ Ⓔ
6. Ⓐ Ⓑ Ⓒ Ⓓ Ⓔ
7. Ⓐ Ⓑ Ⓒ Ⓓ Ⓔ
8. Ⓐ Ⓑ Ⓒ Ⓓ Ⓔ
9. Ⓐ Ⓑ Ⓒ Ⓓ Ⓔ
10. Ⓐ Ⓑ Ⓒ Ⓓ Ⓔ
11. Ⓐ Ⓑ Ⓒ Ⓓ
12. Ⓐ Ⓑ Ⓒ Ⓓ
13. Ⓐ Ⓑ Ⓒ Ⓓ
14. Ⓐ Ⓑ Ⓒ Ⓓ Ⓔ
15. Ⓐ Ⓑ Ⓒ Ⓓ Ⓔ
16. Ⓐ Ⓑ Ⓒ Ⓓ Ⓔ
17. Ⓐ Ⓑ Ⓒ Ⓓ
18. Ⓐ Ⓑ Ⓒ Ⓓ
19. Ⓐ Ⓑ Ⓒ Ⓓ
20. Ⓐ Ⓑ Ⓒ Ⓓ Ⓔ
21. Ⓐ Ⓑ Ⓒ Ⓓ Ⓔ
22. Ⓐ Ⓑ Ⓒ Ⓓ Ⓔ
23. Ⓐ Ⓑ Ⓒ Ⓓ Ⓔ
24. Ⓐ Ⓑ Ⓒ Ⓓ
25. Ⓐ Ⓑ Ⓒ Ⓓ Ⓔ

26. Ⓐ Ⓑ Ⓒ Ⓓ Ⓔ
27. Ⓐ Ⓑ Ⓒ Ⓓ
28. Ⓐ Ⓑ Ⓒ Ⓓ Ⓔ
29. Ⓐ Ⓑ Ⓒ Ⓓ
30. Ⓐ Ⓑ Ⓒ Ⓓ
31. Ⓐ Ⓑ Ⓒ Ⓓ Ⓔ
32. Ⓐ Ⓑ Ⓒ Ⓓ
33. Ⓐ Ⓑ Ⓒ Ⓓ
34. Ⓐ Ⓑ Ⓒ Ⓓ Ⓔ
35. Ⓐ Ⓑ Ⓒ Ⓓ
36. Ⓐ Ⓑ Ⓒ Ⓓ Ⓔ Ⓕ
37. Ⓐ Ⓑ Ⓒ Ⓓ
38. Ⓐ Ⓑ Ⓒ Ⓓ Ⓔ
39. Ⓐ Ⓑ Ⓒ Ⓓ
40. Ⓐ Ⓑ Ⓒ Ⓓ Ⓔ
41. Ⓐ Ⓑ Ⓒ Ⓓ
42. Ⓐ Ⓑ Ⓒ Ⓓ Ⓔ Ⓕ
43. Ⓐ Ⓑ Ⓒ Ⓓ Ⓔ
44. Ⓐ Ⓑ Ⓒ Ⓓ Ⓔ
45. Ⓐ Ⓑ Ⓒ Ⓓ
46. Ⓐ Ⓑ Ⓒ Ⓓ
47. Ⓐ Ⓑ Ⓒ Ⓓ
48. Ⓐ Ⓑ Ⓒ Ⓓ Ⓔ
49. Ⓐ Ⓑ Ⓒ Ⓓ
50. Ⓐ Ⓑ Ⓒ Ⓓ

51. Ⓐ Ⓑ Ⓒ Ⓓ Ⓔ
52. Ⓐ Ⓑ Ⓒ Ⓓ Ⓔ Ⓕ
53. Ⓐ Ⓑ Ⓒ Ⓓ
54. Ⓐ Ⓑ Ⓒ Ⓓ Ⓔ
55. Ⓐ Ⓑ Ⓒ Ⓓ Ⓔ
56. Ⓐ Ⓑ Ⓒ Ⓓ Ⓔ
57. Ⓐ Ⓑ Ⓒ Ⓓ Ⓔ Ⓕ Ⓖ
58. Ⓐ Ⓑ Ⓒ Ⓓ Ⓔ
59. Ⓐ Ⓑ Ⓒ Ⓓ
60. Ⓐ Ⓑ Ⓒ Ⓓ Ⓔ
61. Ⓐ Ⓑ Ⓒ Ⓓ Ⓔ
62. Ⓐ Ⓑ Ⓒ Ⓓ
63. Ⓐ Ⓑ Ⓒ Ⓓ Ⓔ
64. Ⓐ Ⓑ Ⓒ Ⓓ
65. Ⓐ Ⓑ Ⓒ Ⓓ
66. Ⓐ Ⓑ Ⓒ Ⓓ Ⓔ Ⓕ
67. Ⓐ Ⓑ Ⓒ Ⓓ Ⓔ
68. Ⓐ Ⓑ Ⓒ Ⓓ
69. Ⓐ Ⓑ Ⓒ Ⓓ
70. Ⓐ Ⓑ Ⓒ Ⓓ
71. Ⓐ Ⓑ Ⓒ Ⓓ Ⓔ
72. Ⓐ Ⓑ Ⓒ Ⓓ
73. Ⓐ Ⓑ Ⓒ Ⓓ
74. Ⓐ Ⓑ Ⓒ Ⓓ
75. Ⓐ Ⓑ Ⓒ Ⓓ

76. Ⓐ Ⓑ Ⓒ Ⓓ Ⓔ
77. Ⓐ Ⓑ Ⓒ Ⓓ Ⓔ
78. Ⓐ Ⓑ Ⓒ Ⓓ
79. Ⓐ Ⓑ Ⓒ Ⓓ
80. Ⓐ Ⓑ Ⓒ Ⓓ
81. Ⓐ Ⓑ Ⓒ Ⓓ Ⓔ
82. Ⓐ Ⓑ Ⓒ Ⓓ
83. Ⓐ Ⓑ Ⓒ Ⓓ
84. Ⓐ Ⓑ Ⓒ Ⓓ Ⓔ Ⓕ
85. Ⓐ Ⓑ Ⓒ Ⓓ
86. Ⓐ Ⓑ Ⓒ Ⓓ Ⓔ Ⓕ
87. Ⓐ Ⓑ Ⓒ Ⓓ
88. Ⓐ Ⓑ Ⓒ Ⓓ Ⓔ
89. Ⓐ Ⓑ Ⓒ Ⓓ Ⓔ
90. Ⓐ Ⓑ Ⓒ Ⓓ Ⓔ
91. Ⓐ Ⓑ Ⓒ Ⓓ
92. Ⓐ Ⓑ Ⓒ Ⓓ Ⓔ
93. Ⓐ Ⓑ Ⓒ Ⓓ Ⓔ
94. Ⓐ Ⓑ Ⓒ Ⓓ
95. Ⓐ Ⓑ Ⓒ Ⓓ
96. Ⓐ Ⓑ Ⓒ Ⓓ Ⓔ Ⓕ Ⓖ
97. Ⓐ Ⓑ Ⓒ Ⓓ Ⓔ
98. Ⓐ Ⓑ Ⓒ Ⓓ
99. Ⓐ Ⓑ Ⓒ Ⓓ Ⓔ
100. Ⓐ Ⓑ Ⓒ Ⓓ

Practice Exam Part One

1. Who can view CIRs posted to the USGBC website? (Choose two.)

    (A) individuals with a USGBC website account
    (B) registered employees of USGBC member companies
    (C) LEED Accredited Professionals
    (D) registered project team members

2. LEED project teams can earn an "extra" point by achieving _____ performance. (Choose two.)

    (A) exemplary
    (B) ideal
    (C) innovative
    (D) original
    (E) perfect

3. Potable water can also be called _____.

    (A) drinking water
    (B) graywater
    (C) blackwater
    (D) rainwater

4. To increase a LEED project's chances of success, team members should be involved with which of the following project phases? (Choose three.)

    (A) concept
    (B) design development
    (C) inspector site visits
    (D) ongoing commissioning
    (E) construction

5. LEED generally groups credits by credit categories, but the LEED O&M reference guide introduction describes an alternative way of grouping the credits: by their functional characteristics. Which of the following are identified as functional characteristic groups? (Choose two.)

    (A) Administration
    (B) Materials In
    (C) Sustainable Sites
    (D) Waste Management
    (E) Water Efficiency

6. A LEED project is more likely to stay within budget when the team does which of the following? (Choose three.)

    (A) submits documentation for as few LEED credits as possible
    (B) adheres to the plan throughout project
    (C) aligns goals with budget
    (D) contacts USGBC for budget guidance
    (E) establishes project goals and expectations

7. Projects located in urban areas can utilize which of the following to meet LEED open space requirements? (Choose two.)

    (A) accessible roof decks
    (B) landscaping with indigenous plants
    (C) non-vehicular, pedestrian orientated hardscapes
    (D) onsite photovoltaics
    (E) pervious parking lots

8. Which of the following can affect a building's energy efficiency? (Choose three.)

    (A) building orientation
    (B) envelope thermal efficiency
    (C) HVAC system sizing
    (D) refrigerant selection
    (E) VOC content of building materials

9. Which of the following are included in calculations used to determine life-cycle costs? (Choose two.)

    (A) equipment
    (B) facility alterations
    (C) maintenance
    (D) occupant transportation
    (E) utilities

10. Which of the following are NOT common benefits of daylighting? (Choose two.)

    (A) increased productivity
    (B) reduced air pollution
    (C) reduced heat island effect
    (D) reduced light pollution
    (E) reduced operating costs

11. The LEED CI rating system can be applied to which of the following projects? (Choose two.)

    (A) major envelope renovation of a building
    (B) renovation of part of an owner-occupied building
    (C) tenant infill of an existing building
    (D) upgrades to the operation and maintenance of an existing facility

12. Certain prerequisites and credits require project teams to create policies. What information should be included in a policy model? (Choose two.)

    (A) performance period
    (B) policy author
    (C) responsible party
    (D) time period

13. The primary function of which of the following is to encourage sustainable building design, construction, and operation?

    (A) chain-of-custody
    (B) the LEED rating systems
    (C) standard operating procedures (SOPs)
    (D) waste reduction program

14. What information is required to register a project for LEED certification? (Choose three.)

    (A) list of LEED project team members
    (B) name of LEED AP who will be working on project
    (C) primary contact information
    (D) project owner information
    (E) project type

15. Which of the following are among the five basic steps for pursing LEED for Homes certification? (Choose two.)

    (A) achieve certification as a LEED home
    (B) become a USGBC member company
    (C) create list of LEED certified products to be used
    (D) include appliances only if they are Energy Star-rated
    (E) market and sell the home

16. Fire suppression systems that use which of the following will contribute the least to ozone depletion? (Choose two.)

    (A) chloroflourocarbons
    (B) halons
    (C) hydrochlorofluorocarbons
    (D) hydrofluorocarbons
    (E) water

17. Which of the following describes the *property area*?

    (A) area of the project site impacted by the building and hardscapes
    (B) area of the project site impacted by hardscapes only
    (C) area of the project site impacted by the building only
    (D) total area of the site including constructed and non-constructed areas

18. Which LEED rating system includes performance periods and requires recertification to maintain the building's LEED certification?

    (A) LEED CI
    (B) LEED CS
    (C) LEED EBO&M
    (D) LEED NC

19. For a credit uploaded to LEED Online, a white check mark next to a credit name indicates that the credit is _____.

    (A) being pursued and no online documentation has been uploaded
    (B) being pursued and some online documentation has been uploaded
    (C) complete and ready for submission
    (D) not being pursued

20. Remodeling an older existing building may help a project team achieve credit for building reuse and prevent a project team from achieving credit for _____. (Choose three.)

    (A) controllability of systems
    (B) energy use reduction
    (C) regional materials
    (D) sustainable site use
    (E) water use reduction

21. LEED CS (rather than LEED NC) should be pursued when which of the following items are outside the control of the building owner? (Choose three.)

    (A) envelope insulation
    (B) interior finishes
    (C) lighting
    (D) mechanical distribution
    (E) site selection

22. LEED Online provides a means for _____. (Choose two.)

    (A) code officials to access project documentation
    (B) product vendors to advertize
    (C) project team members to analyze anticipated building energy performance
    (D) project administrators to manage LEED projects
    (E) project team members to manage LEED prerequisites and credits

23. Installing a green roof can help a project team achieve which of the following LEED credits? (Choose two.)

    (A) Development Density and Community Connectivity
    (B) Heat Island Effect
    (C) Light Pollution Reduction
    (D) Site Selection
    (E) Stormwater Design

24. Using light bulbs with low mercury content, long life, and high lumen output will result in which of the following?

    (A) improved indoor air quality
    (B) increased light pollution
    (C) reduced light pollution
    (D) reduced toxic waste

25. Which of the following are covered in the Sustainable Sites credit category? (Choose three.)

    (A) light to night sky
    (B) light trespass
    (C) on-site renewable energy
    (D) stormwater mitigation
    (E) refrigerants

26. Which of the following tasks must be part of a durability plan? (Choose two.)

    (A) assign responsibilities for plan implementation
    (B) evaluate durability risks of project
    (C) incorporate durability strategies into design
    (D) research local codes
    (E) submit CIR for available list of durability strategies

27. What is the role of the TSAC? (Choose two.)

    (A) ensure the technical soundness of the LEED reference guides and training
    (B) maintain technical rigor and consistency in the development of LEED credits
    (C) resolve issues to maintain consistency across different LEED rating systems
    (D) respond to CIRs submitted by LEED project teams

28. Project teams can earn credit for purchasing sustainable ongoing consumables within the LEED EBO&M rating system. Ongoing consumables containing a defined amount of _____ are considered sustainable. (Choose three.)

    (A) material certified by the Rainforest Alliance
    (B) preindustrial material
    (C) rapidly renewable material
    (D) regionally extracted material
    (E) salvaged material

29. Within an existing building undergoing major renovations, a tenant is pursuing LEED certification. Which rating system should the tenant use to earn a LEED plaque?

    (A) LEED CI
    (B) LEED CS
    (C) LEED EBO&M
    (D) LEED NC

30. Regional Priority credits vary depending on a project's _____.

    (A) certification level
    (B) community connectivity
    (C) development density
    (D) geographic location

31. Which of the following costs should be considered prior to pursuing a LEED credit? (Choose three.)

    (A) application review cost
    (B) construction cost
    (C) documentation cost
    (D) registration cost
    (E) soft cost

32. Which rating system include CIRs, appeals, and performance periods?

    (A) LEED CI
    (B) LEED CS
    (C) LEED EBO&M
    (D) LEED NC

33. Which standard addresses energy-efficient building design?

    (A) ASHRAE 55-2004
    (B) ANSI/ASHRAE 52.2-1999
    (C) ANSI/ASHRAE 62.1-2007
    (D) ANSI/ASHRAE/IESNA 90.1-2007

34. Which of the following credits will NOT be directly affected by a project team intending to increase a building's ventilation? (Choose two.)

    (A) Controllability of Systems
    (B) Enhanced Commissioning
    (C) Indoor Chemical and Pollutant Source Control
    (D) Measurement and Verification
    (E) Optimize Energy Performance

35. Which of the following describes the purpose of a chain-of-custody document?

    (A) track the movement of products from extraction to the production site
    (B) track the movement of wood products from the forest to the building site
    (C) verify rapidly renewable materials
    (D) verify recycled content

36. A material must be which of the following for it to qualify as a *regional material* within the LEED rating systems? (Choose two.)

    (A) FSC-certified
    (B) made from 10% post-consumer content
    (C) made from products that take 10 years or less to grow
    (D) manufactured within 500 miles of the project site
    (E) permanently installed on the project site
    (F) used to create process equipment

37. A project team's choice of paint will NOT affect which of the following credits?

    (A) Heat Island Effect
    (B) Low-Emitting Materials
    (C) Materials Reuse
    (D) Rapidly Renewable Materials

38. Which of the following do all LEED rating systems contain? (Choose three.)

    (A) core credits
    (B) educational credits
    (C) innovation credits
    (D) operational credits
    (E) prerequisites

39. Which of the following is NOT an example of a durable good?

    (A) computer
    (B) door
    (C) landscaping equipment
    (D) office desk

40. Which of the following would most likely be affected by an increase in ventilation? (Choose two.)

    (A) construction costs
    (B) interior temperature set points
    (C) local climate
    (D) operational costs
    (E) refrigerant management

41. Which items should be considered when selecting refrigerants for a building's HVAC & R system? (Choose two.)

    (A) global depletion potential
    (B) global warming potential
    (C) ozone depletion potential
    (D) ozone warming potential

42. Which information is required to set up a personal account on the USGBC website? (Choose three.)

    (A) company name
    (B) email address
    (C) industry sector
    (D) LEED credentialing exam date
    (E) phone number
    (F) prior LEED projects worked on

43. Project teams must comply with which of the following CIRs? (Choose two.)

    (A) CIRs appealed prior to project registration, for all rating systems
    (B) CIRs reviewed by a TAG for their own project
    (C) CIRs reviewed by a TAG prior to project registration, for projects within the project's climate region
    (D) CIRs posted prior to project application, for the applicable rating system only
    (E) CIRs posted prior to project completion, for projects within their own project's climate region

44. Which of the following describe methods of earning exemplary performance credit? (Choose two.)

    (A) achieve 75% of the credits in each LEED category
    (B) achieve every LEED credit in the Energy and Atmosphere category
    (C) achieve the next incremental level of an existing credit
    (D) double the requirements of an existing credit
    (E) pursue LEED CI certification within a LEED CS-certified building

45. Which of the following potable water conserving strategies may help a project team achieve a LEED point? (Choose two.)

    (A) collecting blackwater for landscape irrigation
    (B) collecting rainwater for sewage conveyance
    (C) installing an on-site septic tank
    (D) using cooling condensate for cooling tower make-up

46. Which of the following requires an explanation of the proposed credit requirements?

    (A) CIR submittal
    (B) Innovation in Design exemplary performance submittal
    (C) LEED credit equivalence submittal
    (D) LEED Online letter template submittal

47. Which of the following organizations defines the off-site renewable energy sources eligible for LEED credits?

    (A) Center for Research and Development of Green Power
    (B) Center for Resource Solutions
    (C) Department of Energy
    (D) Energy Star

48. The International Code Council (ICC) includes which of the following codes? (Choose three.)

    (A) International Building Automation Code (IBAC)
    (B) International Energy Conservation Code (IECC)
    (C) International Lighting Code (ILC)
    (D) International Mechanical Code (IMC)
    (E) International Plumbing Code (IPC)

49. Which of the following is true?

    (A) A LEED project administrator must be a LEED AP.
    (B) Buildings can be LEED-accredited.
    (C) Companies can be USGBC members.
    (D) People can be LEED-certified.

50. In addition to satisfying all prerequisites, what is the minimum percentage of points that a project team can earn to achieve LEED Platinum certification?

    (A) 70%
    (B) 80%
    (C) 90%
    (D) 100%

51. LEED submittal templates provide which of the following? (Choose two.)

    (A) a list of potential strategies to achieve a credit or prerequisite
    (B) a means to modify project documentation
    (C) a means to submit a project for review
    (D) a means to review and submit CIRs
    (E) the project's final scorecard

52. According to the *Sustainable Building Technical Manual*, which of the following are key steps of an environmentally responsive design process? (Choose three.)

    (A) bid
    (B) design
    (C) post-design
    (D) pre-design
    (E) re-bid
    (F) vendor selection

53. Which is true of the LEED CI rating system?

    (A) Precertification allows the owner to market to potential tenants.
    (B) Projects can earn a point for prohibiting smoking within the tenant space.
    (C) Projects can earn half points under the Site Selection credit.
    (D) Projects must recertify every five years to maintain certification status.

54. Once a project undergoes a construction review, what are the potential rulings for each submitted prerequisite and credit? (Choose three.)

    (A) anticipated
    (B) clarify
    (C) deferred
    (D) denied
    (E) earned

55. What is the role of the LEED Steering Committee? (Choose two.)

    (A) delegate responsibility and oversee all LEED committee activities
    (B) develop LEED accreditation exams
    (C) ensure that LEED and its supporting documentation is technically sound
    (D) establish and enforce LEED direction and policy
    (E) respond to CIRs submitted by LEED project teams

56. The LEED Online workspace allows the LEED project administrator to do which of the following? (Choose three.)

    (A) apply for LEED EBO&M precertification
    (B) assign credits to project team members
    (C) build a project team
    (D) review credit appeals submitted by other project teams
    (E) submit projects for review

57. Which of the following items are free? (Choose three.)

    (A) a project's first CIR
    (B) LEED BD & C reference guide
    (C) LEED brochure
    (D) LEED certification
    (E) LEED for Homes rating system
    (F) sample LEED submittal templates
    (G) usgbc.org account

58. What happens if a CIR extends beyond the expertise of the assigned TAG? (Choose two.)

    (A) additional response time may be incurred
    (B) it is rejected
    (C) it is sent to the LEED Steering Committee
    (D) it must be submitted as an Innovation in Design credit
    (E) the TAG will provide the ruling

59. What becomes available once a project is registered?

    (A) LEED project tools
    (B) posted CIRs
    (C) posted credit appeals
    (D) sample submittal templates

60. The LEED rating systems require compliance with which of the following? (Choose two.)

    (A) codes and regulations that address asbestos and water discharge
    (B) codes and regulations that address PCBs and water management
    (C) referenced standards that address fixture performance requirements for water use
    (D) referenced standards that address sustainable forest management practices
    (E) referenced standards that address VOCs

61. Which of the following steps are part of creating an integrated project team? (Choose two.)

    (A) include members from varying industry sectors
    (B) include product vendors in the design phase
    (C) involve a commissioning authority in team members' selection
    (D) involve the LEED AP in design integration
    (E) involve the team in different project phases

62. Facilities undergoing minor alterations and system upgrades must follow which rating system?

    (A) LEED CI
    (B) LEED CS
    (C) LEED EBO&M
    (D) LEED NC

63. Which of the following strategies may help a project team achieve a LEED credit? (Choose two.)

    (A) establish an erosion and sedimentation plan
    (B) establish a location for the storage and collection of recyclables
    (C) install HCFC-based HVAC & R equipment
    (D) use rainwater for sewage conveyance or landscape irrigation
    (E) remediate contaminated soil

64. Which of the following statements is true?

    (A) CIRs are reviewed by a TAG at no additional charge once a project is registered.
    (B) The first step toward LEED certification is passing the Green Associate exam.
    (C) The LEED rating system only applies to commercial buildings.
    (D) To achieve LEED certification, every prerequisite and a minimum number of credits must be achieved.

65. Installing ground source heat pumps would help a project team achieve which of the following credits or prerequisites?

    (A) Green Power
    (B) Optimize Energy Performance
    (C) On-Site Renewable Energy
    (D) Water Use Reduction

66. What are the goals of the Portfolio Program? (Choose two.)

    (A) create a volume accreditation path
    (B) encourage global adoption of sustainable green building practices
    (C) exceed the requirements of ANSI/ASHRAE/IESNA 90.1-2007 by a rating system-designated percentage
    (D) offer a volume certification path
    (E) provide a streamlined certification process for large-scale projects
    (F) provide a template of key data for the design team members to compile

67. When should a LEED project's budget be addressed? (Choose two.)

    (A) before the design phase
    (B) during the construction phase
    (C) during the selection of construction team
    (D) during the selection of design team
    (E) upon project completion

68. Which of the following could help minimize potable water used for the site landscaping and contribute toward earning a LEED landscaping credit?

    (A) designing the site, or the building's roof, with no landscaping
    (B) installing invasive plants
    (C) installing native or adapted plants
    (D) installing turf grass

69. How many Regional Priority points are available for LEED projects?

    (A) 2 points
    (B) 4 points
    (C) 6 points
    (D) 8 points

70. A project team chooses to group credits by functional characteristics. Credits focusing on the measurement of a building's energy performance and ozone protection would be grouped into which of the following categories?

    (A) Energy and Atmosphere
    (B) Energy Metrics
    (C) Materials Out
    (D) Site Management

71. Which of the following are considered principle durability risks? (Choose three.)

    (A) heat islands
    (B) interior moisture loads
    (C) ozone depletion
    (D) pests
    (E) ultraviolet radiation

72. What is the first step of the LEED for Homes certification process?

    (A) become a LEED AP
    (B) contact a LEED for Homes provider
    (C) register with GBCI
    (D) submit a CIR

73. HVAC & R equipment with _____ will contribute to ozone depletion and global warming.

    (A) a manufacture date prior to 2005
    (B) a relatively long equipment life
    (C) minimal refrigerant charge
    (D) refrigerant leakage

74. Sealing ventilation ducts, installing rodent and corrosion proof screens, and using air sealing pump covers are strategies that could be a part of which of the following?

    (A) design charrette
    (B) durability plan
    (C) landscape management plan
    (D) PE exemption form

75. Which standard addresses thermal comfort of building occupants?

    (A) ASHRAE 55-2004
    (B) ANSI/ASHRAE 52.2-1999
    (C) ANSI/ASHRAE 62.1-2007
    (D) ANSI/ASHRAE/IESNA 90.1-2007

76. A LEED project's primary contact must submit which of the following when registering a project? (Choose three.)

    (A) email address
    (B) LEED AP certificate
    (C) LEED project history
    (D) organization name
    (E) individual's title

77. Prior to registering a LEED project, which of the following must be confirmed? (Choose two.)

    (A) necessary CIRs
    (B) precertification
    (C) project cost
    (D) project summary
    (E) project team members

78. Water lost through plant transpiration and evaporation from soil is described by the term _____.

    (A) evapotranspiration
    (B) infiltration
    (C) sublimation
    (D) surface runoff

79. Individuals with a LEED reference guide electronic access code can do which of the following?

    (A) join a LEED project as a team member
    (B) print the LEED reference guide
    (C) purchase a LEED reference guide
    (D) view a protected electronic version of the LEED reference guide

80. LEED credits and prerequisites are presented in a common format in all versions of LEED rating systems. The structure includes which of the following?

    (A) economic impact
    (B) greening opportunities
    (C) intent
    (D) submittal requirements

81. The Materials and Resources category directly addresses which of the following? (Choose two.)

    (A) habitat conservation
    (B) durable goods
    (C) energy consumption
    (D) landscape
    (E) waste stream

82. Building on which of the following sites will most likely have the smallest impact on the environment?

    (A) greenfield
    (B) public parklands
    (C) previously undeveloped site
    (D) urban area

83. LEED submittal templates require which of the following items? (Choose two.)

    (A) declarant's name
    (B) product manufacturer
    (C) project area
    (D) project location

84. The pre-design phase of a LEED project should include which of the following steps? (Choose three.)

    (A) commissioning mechanical systems
    (B) establishing a project budget
    (C) establishing project goals
    (D) site selection
    (E) testing and balancing mechanical systems
    (F) training maintenance staff

85. Which is true about a CIR submittal?

    (A) CIRs must be submitted as text-based inquiries.
    (B) Drawings and specification sheets must be submitted as attachments.
    (C) It must include a complete project narrative.
    (D) Text is limited to 1000 words.

86. Green building design and construction decisions should be guided by which of the following items? (Choose three.)

    (A) bid cost
    (B) construction documents
    (C) design cost
    (D) energy efficiency
    (E) environmental impact
    (F) indoor environment

87. GBCI is a non-profit organization that provides which of the following services? (Choose two.)

    (A) accreditation of industry professionals
    (B) certification of sustainable products
    (C) certification of sustainable buildings
    (D) educational programs on sustainability topics

88. Credits can be earned after which of the following project phases? (Choose two.)

    (A) appeal
    (B) design
    (C) certification
    (D) construction
    (E) post-construction

89. What are the short-term benefits of commissioning? (Choose two.)

    (A) assures credit achievement
    (B) decreases initial project cost
    (C) promotes code compliance
    (D) promotes design efficiency
    (E) reduces design and construction time

90. Conventional fossil-based electricity generation results in which of the following emissions? (Choose three.)

    (A) anthropogenic nitrogen oxide
    (B) carbon dioxide
    (C) carbon monoxide
    (D) sulfur dioxide
    (E) VOCs

91. Which strategy helps minimize a site's heat island effect?

    (A) having a high glazing factor
    (B) installing hardscapes with low SRI values
    (C) maximizing the area of site hardscapes
    (D) shading hardscapes with vegetation

92. The BOD, which includes design information necessary to accomplish the owner's project requirements, must contain which of the following? (Choose three.)

    (A) building materials selection
    (B) indoor environmental quality criteria
    (C) mechanical systems descriptions
    (D) process equipment energy consumption information
    (E) references to applicable codes

93. An integrated project team should include which of the following professionals? (Choose two.)

    (A) code official
    (B) energy sustainability consultant
    (C) landscape architect
    (D) product manufacturer
    (E) utility manager

94. A project is considered a major renovation when at least _____ of the building envelope, interior, or mechanical systems is modified.

    (A) 50%
    (B) 60%
    (C) 70%
    (D) 80%

95. A building that will be partially occupied by the owner may pursue LEED CS certification if the building owner occupies no more than _____ of the building's leasable space.

    (A) 25%
    (B) 50%
    (C) 75%
    (D) 80%

96. The LEED EBO&M rating system includes a Best Management Practices prerequisite. This prerequisite would most likely fall under which of the following credit groupings? (Choose two.)

     (A) Energy and Atmosphere
     (B) Indoor Environmental Quality
     (C) Innovation in Design
     (D) Materials and Resources
     (E) Occupant Health and Productivity
     (F) Operational Effectiveness
     (G) Site Management

97. The Project Details section of the LEED project registration form requires which of the following? (Choose three.)

     (A) company names of all team members
     (B) gross area of the building
     (C) list of likely innovation credits to be pursued
     (D) project budget
     (E) site conditions

98. To be eligible for LEED recertification, a project must be

     (A) precertified under the LEED for Homes rating system
     (B) previously certified under the LEED EBO&M rating system
     (C) previously certified as LEED Platinum under the LEED CS rating system
     (D) previously certified at any level other than LEED Platinum under the LEED NC rating system

99. Which of the following can help reduce a building's energy load? (Choose two.)

     (A) reducing the building's heat island effect
     (B) increasing the ventilation rate
     (C) installing heat recovery systems
     (D) flushing out prior to occupancy
     (E) zoning mechanical systems

100. Minimizing which of the following will improve a building's indoor environmental quality?

     (A) acoustical control
     (B) natural lighting
     (C) ventilation rates
     (D) VOC content in building materials

Practice Exam Part Two

1. (A) (B) (C) (D)
2. (A) (B) (C) (D) (E)
3. (A) (B) (C) (D)
4. (A) (B) (C) (D) (E)
5. (A) (B) (C) (D) (E)
6. (A) (B) (C) (D)
7. (A) (B) (C) (D) (E)
8. (A) (B) (C) (D) (E)
9. (A) (B) (C) (D)
10. (A) (B) (C) (D) (E)
11. (A) (B) (C) (D)
12. (A) (B) (C) (D)
13. (A) (B) (C) (D) (E) (F)
14. (A) (B) (C) (D)
15. (A) (B) (C) (D) (E) (F) (G)
16. (A) (B) (C) (D)
17. (A) (B) (C) (D)
18. (A) (B) (C) (D)
19. (A) (B) (C) (D)
20. (A) (B) (C) (D) (E)
21. (A) (B) (C) (D)
22. (A) (B) (C) (D)
23. (A) (B) (C) (D)
24. (A) (B) (C) (D) (E)
25. (A) (B) (C) (D) (E) (F)
26. (A) (B) (C) (D) (E)
27. (A) (B) (C) (D)
28. (A) (B) (C) (D)
29. (A) (B) (C) (D) (E)
30. (A) (B) (C) (D) (E)
31. (A) (B) (C) (D)
32. (A) (B) (C) (D) (E)
33. (A) (B) (C) (D)
34. (A) (B) (C) (D) (E)
35. (A) (B) (C) (D)
36. (A) (B) (C) (D)
37. (A) (B) (C) (D)
38. (A) (B) (C) (D)
39. (A) (B) (C) (D)
40. (A) (B) (C) (D) (E)
41. (A) (B) (C) (D)
42. (A) (B) (C) (D)
43. (A) (B) (C) (D)
44. (A) (B) (C) (D)
45. (A) (B) (C) (D) (E)
46. (A) (B) (C) (D)
47. (A) (B) (C) (D)
48. (A) (B) (C) (D)
49. (A) (B) (C) (D)
50. (A) (B) (C) (D)
51. (A) (B) (C) (D)
52. (A) (B) (C) (D)
53. (A) (B) (C) (D)
54. (A) (B) (C) (D)
55. (A) (B) (C) (D) (E) (F) (G)
56. (A) (B) (C) (D)
57. (A) (B) (C) (D) (E)
58. (A) (B) (C) (D) (E)
59. (A) (B) (C) (D)
60. (A) (B) (C) (D)
61. (A) (B) (C) (D) (E)
62. (A) (B) (C) (D)
63. (A) (B) (C) (D)
64. (A) (B) (C) (D) (E)
65. (A) (B) (C) (D) (E)
66. (A) (B) (C) (D)
67. (A) (B) (C) (D) (E) (F)
68. (A) (B) (C) (D)
69. (A) (B) (C) (D) (E)
70. (A) (B) (C) (D)
71. (A) (B) (C) (D)
72. (A) (B) (C) (D)
73. (A) (B) (C) (D)
74. (A) (B) (C) (D) (E)
75. (A) (B) (C) (D) (E)

LEED O&M Practice Exam

76. Ⓐ Ⓑ Ⓒ Ⓓ
77. Ⓐ Ⓑ Ⓒ Ⓓ
78. Ⓐ Ⓑ Ⓒ Ⓓ
79. Ⓐ Ⓑ Ⓒ Ⓓ Ⓔ
80. Ⓐ Ⓑ Ⓒ Ⓓ Ⓔ
81. Ⓐ Ⓑ Ⓒ Ⓓ
82. Ⓐ Ⓑ Ⓒ Ⓓ Ⓔ
83. Ⓐ Ⓑ Ⓒ Ⓓ Ⓔ
84. Ⓐ Ⓑ Ⓒ Ⓓ Ⓔ
85. Ⓐ Ⓑ Ⓒ Ⓓ Ⓔ
86. Ⓐ Ⓑ Ⓒ Ⓓ
87. Ⓐ Ⓑ Ⓒ Ⓓ Ⓔ Ⓕ
88. Ⓐ Ⓑ Ⓒ Ⓓ Ⓔ
89. Ⓐ Ⓑ Ⓒ Ⓓ Ⓔ
90. Ⓐ Ⓑ Ⓒ Ⓓ
91. Ⓐ Ⓑ Ⓒ Ⓓ
92. Ⓐ Ⓑ Ⓒ Ⓓ
93. Ⓐ Ⓑ Ⓒ Ⓓ Ⓔ
94. Ⓐ Ⓑ Ⓒ Ⓓ Ⓔ Ⓕ Ⓖ
95. Ⓐ Ⓑ Ⓒ Ⓓ
96. Ⓐ Ⓑ Ⓒ Ⓓ
97. Ⓐ Ⓑ Ⓒ Ⓓ
98. Ⓐ Ⓑ Ⓒ Ⓓ
99. Ⓐ Ⓑ Ⓒ Ⓓ
100. Ⓐ Ⓑ Ⓒ Ⓓ Ⓔ

Practice Exam Part Two

1. A project team has achieved an Energy Star rating of 69, reduced the building occupants' vehicle usage by 25%, purchased 20% off-site renewable energy, and included three LEED APs on the project team. How many points is the team eligible to earn from these strategies?

    (A) 3 points
    (B) 5 points
    (C) 6 points
    (D) 8 points

2. A team registered a project under the LEED EB version 2.0 rating system and began writing the environmental policies, implementing the appropriate green building strategies, and setting up the performance periods after the release of the LEED EBO&M 2009 rating system. The project does not include a major renovation, but the project team is seeking LEED certification of the entire building. Which of the following LEED rating systems could the team be certified under? (Choose two.)

    (A) LEED CI version 2.0 rating system
    (B) LEED EBO&M 2009 rating system
    (C) LEED EBO&M first edition 2008 rating system
    (D) LEED EB version 2.0 rating system
    (E) LEED NC version 2.2 rating system

3. To reduce the negative impact of automobiles on the environment, a building owner is providing 10% of the full-time equivalent (FTE) occupants with free mass transit passes. Using only this strategy, how many points can this project achieve under SS Credit 4, Alternative Commuting Transportation?

    (A) 3 points
    (B) 7 points
    (C) 11 points
    (D) none of the above

4. What is the purpose of documenting the operating costs of a building both before and after implementing a sustainable strategy? (Choose two.)

    (A) determine the financial impacts of various LEED rating system credits
    (B) evaluate the facility manager's ability to operate mechanical systems
    (C) fulfill the requirements of SS Credit 1, LEED Certified Design and Construction
    (D) provide the owner with proof of return on investment for building upgrades
    (E) reduce the carbon footprint of the building

5. A supermarket owner has hired a sustainability consulting firm to create a policy addressing the store's energy requirements. Following an energy audit of the building, the consultants recommend using a building automation system to meter and record the energy use of the HVAC systems. Installing the recommended system might contribute toward which of the following prerequisites or credits? (Choose three.)

    (A) EA Prerequisite 1, Energy Efficiency Best Management Practices
    (B) EA Prerequisite 3, Fundamental Refrigerant Management
    (C) EA Credit 3.1, Performance Measurement: Building Automation System
    (D) EA Credit 3.2, Performance Measurement: System Level Metering
    (E) EA Credit 6, Emissions Reduction Reporting

6. The LEED EBO&M rating system CANNOT be applied to which the following projects?

    (A) commercial buildings with no previous certification
    (B) institutional buildings over 50,000 sq ft
    (C) major renovations of existing buildings
    (D) private high-rise residential buildings

7. The landscape designer of an existing building in a rural area has created a plan to replace unused site hardscapes with native plants. Which of the following measurements must the landscape designer calculate in order to achieve SS Credit 5, Site Disturbance: Protect or Restore Open Habitat. (Choose three.)

    (A) area of the building footprint
    (B) building's total square footage
    (C) parking lot area
    (D) total site area
    (E) vegetated site area

8. IO Credit 1 can be achieved by using which of the following strategies? (Choose two.)

    (A) achieving a LEED established geographically specific credit
    (B) exceeding the requirements of an existing credit by a percentage established by the rating system
    (C) implementing a credit-earning strategy from another LEED rating system
    (D) including a USGBC member on the project team
    (E) including a LEED AP on the project team

9. Which of the following appear on a building's electricity bill?

    (A) greenhouse gas emissions
    (B) on-site renewable energy usage
    (C) site energy usage
    (D) source energy usage

10. Which prerequisites or credits may be achieved by utilizing the online Energy Star Portfolio Manager tool? (Choose three.)

    (A) EA Prerequisite 2, Minimum Energy Efficiency Performance
    (B) EA Credit 1, Optimize Energy Efficiency Performance
    (C) EA Credit 6, Emissions Reduction Reporting
    (D) EQ Prerequisite 1, Minimum IAQ Performance
    (E) EQ Prerequisite 3, Green Cleaning Policy

11. Which of the following CANNOT be registered as a single project under the LEED EBO&M rating system?

    (A) entire buildings that house numerous tenants and occupancies
    (B) individual buildings that are owner or tenant occupied
    (C) individual tenant spaces within an existing building
    (D) multiple buildings owned by the same development company

12. A building's plumbing engineer has designed a system that will collect rainwater and use it to flush toilets. This will reduce the amount of potable water required for sewage conveyance and the amount of stormwater runoff produced from the site. What percentage of the project site's rainwater must the engineer's system collect to achieve SS Credit 6, Stormwater Quantity Control?

    (A) 15% for an average weather year and a one-year, 24-hour design storm
    (B) 15% for an average weather year and a two-year, 24-hour design storm
    (C) 30% for an average weather year and a one-year, 24-hour design storm
    (D) 30% for of an average weather year and a two-year, 24-hour design storm

13. Which of the following are required for a project team to earn a point for its IPM plan? (Choose three).

    (A) minimize human exposure to potential chemical hazards
    (B) plan to notify occupants at least 24 hours before planned pesticide application
    (C) plan to notify occupants no more than 24 hours after emergency pesticide application
    (D) schedule routine inspection and monitoring
    (E) specify circumstances under which emergency pesticide application is acceptable
    (F) specify circumstances under which pesticide application would be chosen over pest removal, and plan to choose pest removal whenever possible

14. Exemption from EA Prerequisite 3, Fundamental Refrigerant Management requires the HVAC & R systems to meet certain limits on either CFC-based refrigerant volume or return on investment. Which of the following describes these limits?

    (A) HVAC & R equipment contains less than 0.5 pounds of refrigerant or the replacement system has a payback of more than 5 years
    (B) HVAC & R equipment contains less than 0.5 pounds of refrigerant or the replacement system has a payback of more than 10 years
    (C) HVAC & R equipment contains less than 5.0 pounds of refrigerant or the replacement system has a payback of more than 5 years
    (D) HVAC & R equipment contains less than 5.0 pounds of refrigerant or the replacement system has a payback of more than 10 years

15. Up to six points are available for implementing which of the following strategies under the LEED EBO&M rating system? (Choose three.)

    (A) alternative commuting transportation
    (B) energy efficiency
    (C) exemplary performance
    (D) green cleaning
    (E) regional priority
    (F) renewable energy
    (G) sustainable purchasing

16. What is the first step required for LEED project certification?

    (A) becoming a corporate member of the USGBC
    (B) having the project administrator become a LEED AP
    (C) registering the project
    (D) submitting a CIR

17. An architectural firm in downtown Chicago has installed a roof garden to provide occupants with a connection to nature. What percentage of the roof area must be vegetated for the team to earn a point for SS Credit 7.2, Heat Island Reduction: Roof?

    (A) 15%
    (B) 30%
    (C) 50%
    (D) 60%

18. Thermal environment is referenced in which of the following standards?

    (A) ANSI/ASHRAE 52.2-1999
    (B) ASHRAE 55-2004
    (C) ANSI/ASHRAE 62.1-2007
    (D) ANSI/ASHRAE/IESNA 90.1-2001

19. A building owner is working with a consultant to reduce the impact of cigarette smoke on the building's indoor air quality. What is the minimum distance required for designated outside smoking areas to be from fresh air intakes, doors, and operable windows, for EQ Prerequisite 2, Environmental Tobacco Smoke (ETS) Control?

    (A) 10 feet
    (B) 15 feet
    (C) 25 feet
    (D) 50 feet

20. MR Prerequisite 2, Solid Waste Management Policy, requires that the building owner, property manager, or facility manager create a policy addressing the building and site's waste. Which of the following credits must the policy address to achieve the prerequisite? (Choose three.)

    (A) MR Credit 5, Sustainable Purchasing: Food
    (B) MR Credit 6, Solid Waste Management: Waste Stream Audit
    (C) MR Credit 7, Solid Waste Management: Ongoing Consumables
    (D) MR Credit 8, Solid Waste Management: Durable Goods
    (E) MR Credit 9, Solid Waste Management: Facility Alterations and Additions

21. After which of the following stages does the project team begin to collect information and perform calculations to meet the requirements of the prerequisites and credits being pursued?

    (A) appeal review
    (B) construction review
    (C) project certification
    (D) project registration

22. Plumbing engineers must follow the UPC or the IPC to earn WE Prerequisite 1, Minimum Indoor Plumbing Fixture and Fitting Efficiency. If the building's year of substantial completion is used to determine the required plumbing efficiency level for this prerequisite. Which combination of substantial completion was before 1993, the baseline for indoor potable water use for this credit will be what percentage of UPC or IPC?

    (A) 60%
    (B) 80%
    (C) 120%
    (D) 160%

23. Increasing lighting control can lead to decreased energy usage and increased productivity. What percentage of workstations and multi-occupant rooms must have with lighting control for a project team to earn a point for lighting control?

    (A) 25%
    (B) 30%
    (C) 40%
    (D) 50%

24. A project's facility manager has been put in charge of creating a waste management plan that focuses on the building's durable goods. Which of the following items should be addressed in the manager's plan? (Choose three.)

    (A) appliances
    (B) mercury-containing light bulbs
    (C) office equipment
    (D) televisions
    (E) toner cartridges

25. If purchased during the performance period, which of the following items contribute to meeting MR Credits 1–5, Sustainable Purchasing? (Choose three.)

    (A) batteries
    (B) carbon credits
    (C) furniture
    (D) graywater
    (E) lamps
    (F) renewable power

26. Which of the following must a green policy include? (Choose three.)

    (A) hand hygiene strategies
    (B) maintenance staff training
    (C) occupancy of facility
    (D) standard operating procedures
    (E) ventilation system schedule

27. Which of the following is NOT a step of the CIR process?

    (A) review the LEED O&M reference guide for direction
    (B) call the CIR hotline for assistance
    (C) review previously posted rulings
    (D) submit a CIR

28. WE Prerequisite 1, Minimum Indoor Plumbing Fixture and Fitting Efficiency, requires establishing the building's baseline potable water usage. How many points may be awarded through WE Credit 2, Additional Indoor Plumbing Fixture and Fitting Efficiency, if the project team achieves a 25% reduction of potable water usage compared to the calculated baseline?

    (A) 1 point
    (B) 3 points
    (C) 4 points
    (D) 5 points

29. A consultant has recommended that a project team pursue MR Credit 6, Solid Waste Management: Waste Stream Audit. Which of the following must be included in the team's waste stream audit to achieve the credit? (Choose three.)

    (A) baseline of waste amounts

    (B) durable goods

    (C) every ongoing consumable

    (D) opportunities to increase recycling and waste diversion

    (E) amount of construction waste generated from alterations

30. A building's existing mechanical ventilating systems have been balanced to meet ANSI/ASHRAE 62.1-2007 ventilation requirements for every occupant. Coupled with permanent monitoring of fresh air rates, this effort may contribute to which of the following prerequisites or credits? (Choose two.)

    (A) EQ Prerequisite 1, Minimum IAQ Performance

    (B) EQ Credit 1.1, IAQ Management Program

    (C) EQ Credit 1.2, Outdoor Air Delivery Monitoring

    (D) EQ Credit 1.3, Increased Ventilation

    (E) EQ Credit 2.1, Occupant Comfort: Occupant Survey

31. A building achieved Gold certification under the LEED CS rating system and every tenant space has achieved Gold certification under the LEED CI rating system. Which rating system should the building owner follow to add a Platinum plaque to the collection, and will this be considered an initial certification or recertification?

    (A) LEED CI recertification

    (B) LEED NC initial certification

    (C) LEED EBO&M initial certification

    (D) LEED EBO&M recertification

32. A building operating plan should include which of the following? (Choose three.)

    (A) basis of design

    (B) design indoor conditions

    (C) mode of operation

    (D) owner's project requirements

    (E) time-of-day schedules

33. At least what percent of the building site area must be vegetated for WE Credit 3, Water Efficient Landscaping, and how many points are available through this credit?

    (A) 5%, 3 points

    (B) 5%, 5 points

    (C) 10%, 3 points

    (D) 10%, 5 points

34. Which of the following are considered on-site renewable energy sources under the LEED EBO&M rating system? (Choose two.)

    (A) architectural shading features
    (B) ground source heat pumps
    (C) hydroelectricity
    (D) photovoltaic panels
    (E) windmills

35. A landscaping contractor has installed native vegetation on more than 50% of a roof's surface area to reduce the building's heat island effect and mitigate stormwater run-off. Excluding the opportunity to earn IO points, how many points may this strategy achieve?

    (A) 1 point
    (B) 2 points
    (C) 3 points
    (D) 5 points

36. Which of the following performance period calculations must be compared to that of the previous five years to achieve IO Credit 3, Documenting Sustainable Building Cost Impacts?

    (A) building energy consumption costs
    (B) building operating costs
    (C) occupant productivity
    (D) sustainable consulting costs

37. For a project to be certified under the LEED EBO&M rating system it must comply with all applicable building codes, have at least _____ of the floor area occupied for at least _____ prior to submittal of the certification application, and may exclude up to _____ of the floor area if its operations are under separate management.

    (A) 75%, 6 months, 5%
    (B) 75%, 12 months, 10%
    (C) 85%, 6 months, 10%
    (D) 85%, 12 months, 5%

38. A research and development company installed photovoltaic panels to generate on-site energy and monitored them for a year after installation. With 12% of the building's energy provided from the photovoltaic system, how many points can the team earn under EA Credit 4, Renewable Energy?

    (A) 1 point
    (B) 3 points
    (C) 5 points
    (D) 6 points

39. Which of the following prerequisites and credits can be achieved by complying with the Montreal Protocol?

   (A) EA Prerequisite 1, Energy Efficiency Best Management Practices
   (B) EA Credit 5, Enhanced Refrigerant Management
   (C) EA Credit 6, Emissions Reduction Reporting
   (D) EQ Prerequisite 1, Minimum Indoor Air Quality Performance

40. The owner of a marketing company intends to comply with EQ Credit 3.2, Green Cleaning: Custodial Effectiveness Assessment, because it is proven that occupants have fewer health problems in well cleaned and well maintained facilities. Who can be the declarant for this credit? (Choose two.)

   (A) facility manager
   (B) LEED project administrator
   (C) owner
   (D) property manager
   (E) tenant

41. Areas affected by project site activity are defined as the _____.

   (A) building footprint
   (B) development footprint
   (C) project boundary
   (D) property boundary

42. Installing low-reflectance surfaces can help a project team do which of the following?

   (A) minimize heat islands
   (B) minimize light pollution
   (C) optimize daylight and views
   (D) optimize energy efficiency

43. How often can a project recertify its LEED certification status under the EBO&M rating system?

   (A) as often as every 2 years and at least every 5 years
   (B) as often as every 6 months and at least every 2 years
   (C) as often as every year and at least every 3 years
   (D) as often as every year and at least every 5 years

44. A company's purchasing department is currently drafting a program for environmentally friendly purchasing. To comply with MR Credit 4, Sustainable Purchasing: Reduced Mercury in Lamps, at least _____ of all lights must average a maximum of _____ of mercury per lumen-hour.

    (A) 50%, 80 picograms
    (B) 50%, 90 picograms
    (C) 90%, 60 picograms
    (D) 90%, 90 picograms

45. A contractor has installed a building automation system that controls the HVAC & R and lighting systems. This system can contribute toward achieving which of the following credits? (Choose two.)

    (A) EA Credit 1, Optimize Energy Efficiency Performance
    (B) EA Credit 2, Existing Building Commissioning
    (C) EA Credit 3.1, Performance Measurement: Building Automation System
    (D) EA Credit 5, Enhanced Refrigerant Management
    (E) EQ Credit 2.2, Controllability of Systems: Lighting

46. Which of the following terms is defined as a tracking procedure for documenting the status of a product from the point of harvest or extraction to the ultimate consumer end use, including all stages of processing, transformation, manufacturing, and distribution?

    (A) chain-of-custody
    (B) churn
    (C) energy audit
    (D) fairtrade

47. What is the definition of a *performance period*?

    (A) the time during which operations performance is measured
    (B) the period between project registration and project certification
    (C) the period when the LEED project team must submit CIRs and appeals
    (D) the time the design team spends writing policies and strategies for LEED certification

48. A large manufacturing facility is pursuing Gold certification under the LEED EBO&M rating system. The cafeteria supervisor designing the food purchasing plan has decided to comply with MR Credit 5, Sustainable Purchasing: Food using the LEED-established strategy of buying food produced within 100-miles of the site. What percentage of the total food cost during the performance period must be from these local producers?

    (A) 10%
    (B) 15%
    (C) 20%
    (D) 25%

49. How many points can be earned under SS Credit 1, LEED Certified Design and Construction, for buildings previously certified under the LEED NC rating system?

    (A) 1 point
    (B) 2 points
    (C) 3 points
    (D) 4 points

50. Which of the following is defined as the person who oversees operations, maintenance, and upkeep of the building and serves as a liaison between the owner and the tenants?

    (A) building engineer
    (B) facility manager
    (C) general manager
    (D) property manager

51. Which of the following is NOT one of the levels of energy audits that ASHRAE performs?

    (A) energy survey and analysis
    (B) investment-grade audit
    (C) utility bill analysis
    (D) walk through analysis

52. A building owner has required that a commissioning authority provide ongoing commissioning. If a LEED credit is achieved, what is the maximum length of the commissioning cycle?

    (A) 12 months
    (B) 24 months
    (C) 36 months
    (D) 48 months

53. LEED certification under the EBO&M rating system requires that all performance periods overlap and end within _____ of each other. A project team must complete the prerequisite and credit submittals within _____ of the end of the performance periods.

    (A) 7 days, 60 days
    (B) 30 days, 60 days
    (C) 30 days, 90 days
    (D) 60 days, 90 days

LEED O&M Practice Exam

54. A project team has created a program to minimize its impact on landfills and promote its dedication to sustainability. What percentage of the durable goods must the team divert from the waste stream during the performance period to earn one point under MR Credit 8, Solid Waste Management: Durable Goods?

    (A) 25%
    (B) 50%
    (C) 75%
    (D) 90%

55. A project team is pursuing SS Credit 2, Building Exterior and Hardscape Management Plan. Which of the following items should be addressed in the plan? (Choose two.)

    (A) cleaning agents for site hardscapes
    (B) interior asbestos
    (C) lawn mowers
    (D) pavement coatings
    (E) recycling locations
    (F) roof drainage system
    (G) snow removal chemicals

56. The LEED AP must be which of the following for a project team to earn a point for IO Credit 2, LEED Accredited Professional?

    (A) accredited under the same rating system and version as the project
    (B) a member of USGBC
    (C) a principle participant on the project team
    (D) the declarant of at least 60% of the LEED submittal templates

57. Which of the following must be described in a project narrative for LEED certification under the EBO&M rating system? (Choose three.)

    (A) building and site
    (B) building history
    (C) building occupancy and use
    (D) project challenges
    (E) project summary and scope

58. A janitorial service provides a green cleaning program that reduces the environmental impact of cleaning products. Which of the following products are included in the 30% that must meet the required sustainable criteria to achieve one LEED point for EQ Credit 3.3, Green Cleaning: Purchase of Sustainable Cleaning Products and Materials?

    (A) air cleaning devices
    (B) carpet care products
    (C) cleaning equipment
    (D) sealants

59. A law firm owner would like to improve the indoor air quality throughout the building. Following examination of the existing mechanical ventilation systems, the HVAC consultant determines that the building occupants would benefit from additional fresh air. To achieve EQ Credit 1.3, Increased Ventilation, the building's ventilation rates must exceed the ANSI/ASHRAE 62.1-2007 requirements by what percentage?

(A) 10%
(B) 20%
(C) 30%
(D) 40%

60. Which of the following is a tool that project teams can use to calculate the energy performance of an existing facility?

(A) ANSI/ASHRAE/IESNA 90.1-2007, Appendix G
(B) Energy Star Benchmark
(C) Energy Star Portfolio Manager
(D) LEED Online submittal template

61. Which of the following are intents of SS Credit 3, Integrated Pest Management, Erosion Control and Landscape Management Plan? (Choose three.)

(A) eliminate light trespass from the building site
(B) enhance natural diversity and protect wildlife
(C) integrate high-performance building operations into the surrounding landscape
(D) preserve ecological integrity
(E) reduce the heat island effect resulting from site hardscapes

62. A project pursuing EA Credit 4, Renewable Energy, can utilize the Energy Star Portfolio Manager tool to determine which of the following? (Choose two.)

(A) the number of off-site renewable energy certificates to be purchased
(B) the number of on-site renewable energy certificates to be generated
(C) the percentage of off-site renewable energy certificates to be purchased
(D) the percentage of on-site renewable energy certificates to be generated

63. The LEED O&M reference guide uses the term *usage group* to refer to a group of people within a building that uses a percentage of which of the following?

(A) bicycle racks
(B) preferred parking
(C) ventilation airflow
(D) washroom facilities

64. What must be continuously monitored to achieve EQ Credit 2.3, Occupant Comfort: Thermal Comfort Monitoring? (Choose two.)

   (A) air speed
   (B) air temperature
   (C) ambient noise
   (D) humidity
   (E) radiant temperature

65. In which of the following situations should an existing building pursue certification under the LEED NC rating system rather than the LEED EBO&M rating system? (Choose three.)

   (A) when more than 50% of the occupancy will change
   (B) when alterations will affect more than 50% of the building's floor area
   (C) when more than 50% of the building's occupants will relocate due to alterations
   (D) when the total building area will be expanded by more than 50%
   (E) when alterations will affect more than 50% of the project site

66. A retail store owner is pursuing LEED certification under the EBO&M rating system. The owner intends to increase the store's air quality by pursuing EQ Credit 1.4, Reduce Particulates in Air Distribution. For this credit the HVAC contractor will have to document the planned filter replacement schedule and install a filtration system meeting which of the following minimum specifications?

   (A) MERV 8 filters filtering 100% of the outside air and 50% of the return air
   (B) MERV 8 filters filtering 100% of the outside air and 100% of the return air
   (C) MERV 13 filters filtering 100% of the outside air and 50% of the return air
   (D) MERV 13 filters filtering 100% of the outside air and 100% of the return air

67. Reducing the potable water use of which of the following can help achieve Water Efficiency credits? (Choose three.)

   (A) clothes washers
   (B) cooling tower make-up
   (C) drinking fountains
   (D) dishwashers
   (E) landscaping
   (F) showers and faucets

68. An architecture firm has 200 full-time employees. It does not currently employ interns or part-time employees. The installed water closets use one gallon of water per flush and the urinals use a half gallon of water per flush. Given the default ratio of men to women and "normal" water closet and urinal use, approximately how much water do these fixtures use in one occupied day?

    (A) 300 gallons
    (B) 400 gallons
    (C) 500 gallons
    (D) none of the above

69. Which of the following team members can be the declarant for WE Credit 4, Cooling Tower Water Management? (Choose three.)

    (A) building engineer
    (B) building owner
    (C) facility manager
    (D) groundskeeper
    (E) property manager

70. SS Credit 8, Light Pollution Reduction, offers a point for environmentally responsible interior and exterior lighting design. Which of the following exterior lighting strategies meets the requirements of this credit?

    (A) install automatic sensors to turn off all interior lighting after dusk
    (B) remove all exterior lighting 100 watts and over
    (C) shield all exterior light fixtures 50 watts and over
    (D) utilize solar-powered exterior lighting

71. MR Prerequisite 1, Sustainable Purchasing Policy, requires that the project team create a product purchasing policy to implement the other MR Sustainable Purchasing credits. The policy can address which combination of the following credits to comply with MR Prerequisite 1?

    (A) durable goods and food credits only
    (B) durable goods and facility alterations and additions credits only
    (C) ongoing consumables and reduced mercury in lamps credits only
    (D) ongoing consumables and food credits only

72. What is the maximum number of points a project team can earn for IO Credit 1?

    (A) 2 points
    (B) 3 points
    (C) 4 points
    (D) 6 points

73. Installing which of the following will NOT contribute toward meeting the requirement of SS Credit 5, Site Development: Protect or Restore Open Habitat?

    (A) adapted vegetation off site equal to at least 50% of the project site area
    (B) a vegetated roof equal to at least 5% of the project site area
    (C) turf grass on site equal to at least 25% of the project site area
    (D) native vegetation on site equal to 25% of the project site area

74. Furniture is considered a durable good and can help a project team earn points if purchased during the performance period and if the furniture _____. (Choose three.)

    (A) contains no added urea-formaldehyde
    (B) is 70% salvaged
    (C) is made of 50% rapidly renewable material
    (D) is made of at least 50% FSC-certified wood
    (E) uses no VOCs in its adhesives

75. Once a project is registered the project team gains access to which of the following tools? (Choose three.)

    (A) exemplary credits available
    (B) minimum narrative requirements
    (C) offline credit calculators
    (D) policy, program, and plan models
    (E) member-to-member exchange

76. How does the LEED O&M reference guide define *daylighting*?

    (A) controlled modifications in the lighting levels during the day hours
    (B) lighting that requires a setback cycle during night time hours
    (C) transmission of high spectrum light into the space through Energy Star light bulbs
    (D) transmission of natural light into a space to reduce electric lighting

77. The condensation from a building's air conditioning system is collected, stored, and used as first stage make-up water for the cooling tower loop. The system provides a payback to the owner because it reduces the need for potable water. How much of the make-up water must come from a non-potable source to achieve WE Credit 4.2, Cooling Tower Water Management: Non-Potable Water Source Use?

    (A) 10%
    (B) 25%
    (C) 50%
    (D) 75%

Practice Exam Part Two

78. A convention center meeting ANSI/ASHRAE 62.1-2007 ventilation standards is registered for LEED certification under the EBO&M rating system. The project's HVAC consultant proposes installing energy recovery ventilators to recover some the energy lost from the relief air stream. In addition to EQ Credit 1.3, Increased Ventilation, which of the following credits or prerequisites will be directly affected by this strategy?

    (A) SS Credit 1, LEED Certified Design and Construction
    (B) EA Prerequisite 3, Fundamental Refrigerant Management
    (C) EA Credit 1, Optimize Energy Efficiency Performance
    (D) EA Credit 4, Renewable Energy

79. SMACNA's *IAQ Guidelines for Occupied Buildings Under Construction* recommends that control measures be implemented to address which of the following areas? (Choose three.)

    (A) cleaning agents
    (B) HVAC & R protection
    (C) pathway interruption
    (D) source control
    (E) VOC Limits

80. Projects previously certified under other LEED rating systems and now pursuing certification under the LEED EBO&M rating system are eligible to utilize a streamlined path for online submittal. Which credits are available for this path? (Choose two.)

    (A) SS Credit 1, LEED Certified Design and Construction
    (B) SS Credit 3, Integrated Pest Management, Erosion Control, and Landscape Management Plan
    (C) EA Credit 5, Enhanced Refrigerant Management
    (D) MR Credit 5, Sustainable Purchasing: Food
    (E) MR Credit 7, Solid Waste Management: Ongoing Consumables

81. An office manager would like to optimize thermal conditions to promote office productivity. What is the minimum percentage of occupants that must complete the thermal comfort survey for EQ Credit 2.1, Occupant Comfort: Occupant Survey?

    (A) 30% of the full-time occupants and 20% of the transient occupants
    (B) 50% of the full-time occupants and 30% of the transient occupants
    (C) 30% of the full-time occupants
    (D) 50% of all occupants

82. Which of the following commissioning steps are addressed by EA credits? (Choose three.)

    (A) documentation
    (B) investigation and analysis
    (C) implementation
    (D) ongoing commissioning
    (E) replacement

83. A landscape architect is designing a site to reduce the amount of potable water used for landscaping and achieve WE Credit 3, Water Efficient Landscaping. If he is not interested in xeriscaping, which of the following must he do for his design to earn points for this credit? (Choose three.)

    (A) calculate the landscaped area of site
    (B) estimate the annual rainfall of site location
    (C) estimate the available blackwater for irrigation
    (D) estimate the available graywater for irrigation
    (E) meter potable water used for irrigation

84. Which of the following people can be the declarant for SS Credit 2, Building Exterior and Hardscape Management Plan? (Choose three.)

    (A) civil engineer
    (B) facility manager
    (C) groundskeeper
    (D) landscaper
    (E) property manager

85. The LEED EBO&M project team can choose to follow a streamlined path in lieu of the standard submittal path by utilizing the Licensed Professional Exemption Form. Which project team members can provide the required documentation on this form? (Choose three.)

    (A) administrator
    (B) commissioning authority
    (C) registered architect
    (D) registered landscape architect
    (E) registered professional engineer

86. A permanent ventilation monitoring system will verify that a building's air handlers are providing fresh air at the ANSI/ASHRAE 62.1-2007 required rates during building occupation. An alarm is generated if the carbon dioxide rate rises 15% above the code requirement. Given these conditions, how many of the building's ten air handlers must have an outside air monitoring station to achieve a point in the LEED EBO&M rating system?

    (A) one of the air handlers
    (B) five of the air handlers
    (C) eight of the air handlers
    (D) all of the air handlers

87. Which of the following installations will contribute to WE Credit 2, Additional Indoor Plumbing Fixture and Fitting Efficiency? (Choose three.)

    (A) aerators in faucets
    (B) conductivity meters
    (C) drift eliminators
    (D) dual flush water closets
    (E) low-flow showers
    (F) irrigation meters

88. For a retail facility, which of the following people are included in the FTE calculation? (Choose three.)

    (A) customers
    (B) full-time employees
    (C) landscape maintenance crew
    (D) part-time employees
    (E) students

89. Which of the following steps are part of the CIR process? (Choose two.)

    (A) the project team reviews rulings from all previous and current LEED rating systems
    (B) the project team submits a CIR online with the certification application
    (C) CIRs are collected twice a month
    (D) the TAG contacts the LEED project team administrator with the ruling
    (E) rulings are posted online within 12 business days of collection

90. A warehouse is pursuing LEED certification under the EBO&M rating system and is using the Energy Star's Portfolio Manager tool to generate its energy performance rating. What is the minimum EPA Energy Star rating required for the warehouse to be eligible for LEED certification under the EBO&M rating system?

    (A) 50
    (B) 69
    (C) 71
    (D) 75

91. A manufacturing plant's facility manager would like to minimize the thermal gradient difference between the site's parking lot and landscaped areas. The parking lot will be covered with a concrete, non-vegetated structure. To comply with option 2 of SS Credit 7.1, Heat Island Reduction: Non-Roof, at least _____ of the parking must be covered, the cover and exposed portion of the lot must have a _____ SRI of 29, and reflective surfaces must be cleaned _____.

    (A) 50%, maximum, yearly
    (B) 50%, minimum, every other year
    (C) 75%, maximum, every other year
    (D) 75%, minimum, yearly

92. For the purposes of LEED credit documentation, greenhouse gas emissions are recorded in _____.

    (A) pounds of carbon calcium
    (B) pounds of carbonite
    (C) tons of carbon dioxide
    (D) tons of carbon monoxide

93. Which of the following organizations' food certifications will contribute toward MR Credit 5, Sustainable Purchasing: Food? (Choose three.)

    (A) Food Alliance
    (B) Food and Drug Administration
    (C) Marine Stewardship Council
    (D) National Organic Coalition
    (E) Protected Harvest

94. Which of the following influence the fees a LEED EBO&M project pays for registration and certification application? (Choose three.)

    (A) achieving Platinum level LEED certification
    (B) having a LEED AP on the project team
    (C) having a USGBC company member register the project
    (D) registering the project online
    (E) the building's floor area
    (F) the number of CIRs submitted
    (G) the project's location

95. Mats are installed inside the vestibules of all of a building's public entryways to prevent the entry of particulates on the occupants' shoes. How long (in the direction of travel) must the mats be to comply with EQ Credit 3.5, Indoor Chemical and Pollutant Source Control?

    (A) 5 feet
    (B) 6 feet
    (C) 10 feet
    (D) the entire length of the vestibule

96. To comply with WE Credit 1.2, Water Performance Measurement: Submetering, the installed water meters can measure which of the following?

    (A) all of the cooling tower make-up water
    (B) graywater used for landscaping
    (C) rainwater used for urinals and water closets
    (D) water usage of at least 50% all the indoor plumbing fixtures and fittings

97. An owner is working with an architect and a lighting designer to achieve EQ Credit 2.4, Daylight and Views by utilizing natural light to illuminate the interior of the building. To achieve this credit, the project team must document both the floor area of the _____, and calculate the daylight illumination levels taken on a _____ grid.

    (A) entire building, 10-foot
    (B) entire building, 20-foot
    (C) regularly occupied spaces, 10-foot
    (D) regularly occupied spaces, 20-foot

98. How many points are needed to achieve LEED Silver certification?

    (A) 40 points
    (B) 49 points
    (C) 50 points
    (D) 59 points

99. How much fresh air per occupant must be provided if the ventilation rates of ANSI/ASHRAE 62.1-2007 cannot be met?

    (A) 5 cu ft per minute
    (B) 10 cu ft per minute
    (C) 15 cu ft per minute
    (D) 20 cu ft per minute

100. A mechanical contractor has installed a conductivity meter to maintain proper concentrations of chemicals in the cooling tower water. The contractor has also written a cooling tower make-up water management plan addressing chemical treatment, bleed-off, biological control, and required maintenance of the installed system to comply with WE Credit 4.1, Cooling Tower Water Management: Chemical Management. Who can be the declarant for this credit? (Choose three.)

    (A) building engineer
    (B) commissioning authority
    (C) facility manager
    (D) mechanical contractor
    (E) property manager

Practice Exam Part One Solutions

1. B
2. A
3. A
4. A, B
5. A, B
6. B, C, E
7. A, C
8. A, B, C
9. C
10. C, D
11. B, C
12. C, D
13. B
14. C, D, E
15. A
16. D, E
17. D, E
18. C, E
19. B
20. B, D, E
21. B, C, D
22. D, E
23. A
24. D, E
25. A, B, D
26. B, C
27. A, C, E
28. B, C, D
29. A
30. D, E
31. B, C
32. C
33. D, E
34. A, C
35. B
36. D, E
37. C
38. A, C, E
39. B
40. C, D
41. B, C
42. B, C, E
43. B, D
44. C, D
45. D, E
46. C
47. B, C, E
48. B, D, E
49. C
50. B
51. C
52. A, B, D
53. C, E
54. B, D, E
55. A, D
56. B, C, D, E
57. E, F, G
58. A, C, D
59. A, D, E
60. A, B
61. A, E
62. E
63. D, E
64. D, E
65. B, C, D, E
66. D
67. A, B
68. D, E
69. B
70. B, D, E
71. B, D, E
72. C, D, E
73. D
74. B
75. A

LEED O&M Practice Exam

76. **A** B C **D** **E**
77. A B C **D** **E**
78. **A** B C D
79. A B C **D**
80. A B **C** D
81. A **B** C D **E**
82. A B C **D**
83. **A** B C D
84. A **B** **C** E F
85. **A** B C D
86. A B C **D** **E**
87. **A** B **C** D
88. **A** B C **D** E
89. A B C **D** **E**

90. **A** **B** C **D** E
91. A B C **D**
92. A **B** **C** D **E**
93. A **B** **C** D E
94. **A** B C D
95. A **B** C D
96. **A** B C D E **F** G
97. A **B** C **D** **E**
98. A **B** C D
99. A B **C** D **E**
100. A B C **D**

Practice Exam Part One Solutions

1. *The answers are:* (B) registered employees of USGBC member companies
   (D) registered project team members

Only members of a LEED project team and employees of a USGBC member company may view posted CIRs.

2. *The answers are:* (A) exemplary
   (C) innovative

To earn points in the Innovation in Design category, project teams can exceed the requirements of an existing LEED credit and earn exemplary performance points, or they can implement innovative performance strategies not addressed by the LEED rating systems.

3. *The answer is:* (A) drinking water

Potable water is water that meets or exceeds EPA drinking water standards and is supplied from wells or municipal water systems.

4. *The answers are:* (A) concept
   (B) design development
   (E) construction

LEED projects benefit from the inclusion of project team members in ongoing commissioning and inspector site visits; however, team member participation is most valuable during the design and construction phases of the project.

5. *The answers are:* (A) Administration
   (B) Materials In

The functional characteristic groups established under the EBO&M rating system are Materials In (includes credits addressing the sustainable purchasing policy of a building), Materials Out, Administration (includes credits addressing the planning and logistics support of operating a high-performance building), Green Cleaning, Site Management, Occupational Health and Productivity, Energy Metrics, and Operational Effectiveness. Sustainable Sites and Water Efficiency are credit categories, not functional characteristic groups. Waste Management is neither a credit category nor a functional characteristic group.

6. *The answers are:* (B) adheres to the plan throughout project
   (C) aligns goals with budget
   (E) establishes project goals and expectations

A project is more likely to stay within budget when the goals and budget are coordinated and when the project team adheres to the plan and frequently checks expenses against the budget. USGBC does not provide guidance for project budgeting. Submitting documentation for fewer credits does not necessarily lead to a successful LEED project budget.

7. *The answers are:* (A) accessible roof decks
   (C) non-vehicular, pedestrian orientated hardscapes

Projects located in urban areas (those buildings with little or no setback) can utilize pedestrian hardscapes, pocket parks, accessible roof decks, plazas, and courtyards to meet the open space requirements.

On-site photovoltaics contribute to on-site renewable energy generation. Pervious parking lots contribute to on-site stormwater mitigation. Landscaping with indigenous plants may reduce the amount of water needed for irrigation. None of these strategies contribute to open space requirements.

8. *The answers are:* (A) building orientation
(B) envelope thermal efficiency
(C) HVAC system sizing

A building's overall energy consumption can vary depending on building orientation, building location, envelope insulation, fenestration u-values, the size of the HVAC system, and electricity requirements of lighting and appliances. Addressing these factors during the design can result in reduced energy bills for the life of the building. Refrigerant selection affects the building's environmental impact (but not the HVAC system efficiency), and volatile organic compound (VOC) content affects indoor air quality.

9. *The answers are:* (C) maintenance
(E) utilities

Life-cycle costing is an accounting methodology used to evaluate the economic performance of a product or system over its useful life. Life-cycle cost calculations include maintenance and operating costs (including the cost of utilities). Neither the occupants' individual expenses nor the initial cost of an investment factor into a building's life-cycle costs.

10. *The answers are:* (C) reduced heat island effect
(D) reduced light pollution

The primary purpose of increasing a building's daylighting is to reduce the need for electric light. Reducing electric light use results in reduced electricity use in general, thereby reducing energy costs and the carbon dioxide emissions (or air pollution) created by the building. Additionally, statistics show that productivity is markedly improved in day lit buildings compared to buildings that rely heavily on electric lighting.

Light pollution is related to the amount of light transmitted from a building afterhours. Daylighting is related to daytime lighting of the interior of the building. A building's heat islands are not affected by daylighting.

11. *The answers are:* (B) renovation of part of an owner-occupied building
(C) tenant infill of an existing building

The LEED for Commercial Interiors (CI) rating system provides the opportunity for tenant spaces and parts of buildings to achieve LEED certification. Major building envelope renovations would be certified under the LEED for New Construction (NC) rating system. Upgrades to the operations and maintenance of an existing facility would be certified under the LEED for Existing Buildings: Operations & Maintenance (EBO&M) rating system.

Practice Exam Part One Solutions

12. *The answers are:* (C) responsible party
    (D) time period

Each policy should identify the individual or team responsible its implementation. Additionally, the time period over which the policy is applicable should be indentified. The time period is not necessarily the same amount of time as the performance period. Note that performance periods only apply to LEED EBO&M projects.

13. *The answer is:* (B) the LEED rating systems

Chain-of-custody is a procedure to document the status of a product from the point-of-harvest or extraction to the ultimate consumer end use and can promote sustainable construction. Standard operating procedures (SOPs) are detailed, written instructions that document a method with the intention of achieving uniformity. A waste reduction program helps a project team minimize waste by using source reduction, reuse, and recycling. Both SOPs and waste reduction programs promote sustainable operations. Of the answer options, only the LEED rating systems support design, construction, and operations.

14. *The answers are:* (C) primary contact information
    (D) project owner information
    (E) project type

To register a project for LEED certification, the registrant must provide account login information, primary contact information, project owner information, general project information, payment information, and the project type. Project team member names do not have to be submitted to complete the registration form. LEED projects are not required to have a LEED AP on the team.

15. *The answers are:* (A) achieve certification as a LEED home
    (E) market and sell the home

The five basic steps of the LEED for Homes are

1. contact a LEED for Homes Certification provider and join the program
2. identify the project team
3. build the home to the stated goals
4. achieve certification as a LEED home
5. market and sell the home

Becoming a USGBC member company reduces the registration fee; however, it is not a basic step.

16. *The answers are:* (D) hydrofluorocarbons
    (E) water

Using halons, chlorofluorocarbons, and hydrochlorofluorocarbons in fire suppression systems can lead to ozone depletion. Using water and hydroflourocarbons will have a minimal effect on ozone depletion.

17. The answer is: (D) total area of the site including constructed and non-constructed areas

The total area within the legal property boundaries of the site is considered the property area. The project site area that includes constructed and non-constructed areas is defined as the development footprint.

18. The answer is: (C) LEED EBO&M

LEED EBO&M is the only rating system that includes performance periods and a recertification requirement. LEED EBO&M projects may recertify as often as every year, and must recertify at least every five years to maintain their LEED certification status.

19. The answer is: (B) being pursued and some online documentation has been uploaded

A white check mark indicates that the credit is being pursued and has been assigned to a project member. Some documentation has been uploaded; however, additional info needs to be submitted.

20. The answers are: (B) energy use reduction
(D) sustainable site use
(E) water use reduction

Remodeled projects typically achieve Building Reuse credits within the Materials and Resources credit category; however, they may have difficulty achieving credits in the Sustainable Sites, Water Efficiency, and Energy and Atmosphere categories. This is because when the site is predetermined, design teams do not have the opportunity to select a more favorable site that would help them achieve those credits. Compared to newer buildings, older buildings usually have less energy-efficient insulation systems, and have less efficient plumbing fixtures.

21. The answers are: (B) interior finishes
(C) lighting
(D) mechanical distribution

The LEED Core & Shell (CS) rating system is designed to help designers, builders, developers, and new building owners increase the sustainability of a new building's core and shell construction. It covers base building elements and complement the LEED for Commercial Interiors (CI) rating system. Interior space layout, interior finishes, lighting, and mechanical distribution may not be directly controlled by the developer, and therefore if a project team wishes to include these elements in their LEED certification, LEED CS may not be appropriate. Site selection and the building's envelope insulation can be directly controlled by the owner when pursuing LEED CS.

22. The answers are: (D) project administrators to manage LEED projects
(E) project team members to manage LEED prerequisites and credits

LEED Online does not provide advertising of any sort. Project administrators assign access responsibilities for prerequisites and credits to project team members using LEED Online. Local code officials are unable to view online LEED documentation. The Energy Star Target Finder tool will help a project team analyze anticipated building energy performance.

23. *The answers are:* (A) Heat Island Effect
    (E) Stormwater Design

Green roofs help mitigate stormwater and reduce the roof's heat island effect by increasing evapotranspiration (which has a cooling affect), and increasing the roof's albedo. Light pollution reduction is achieved through strategic lighting design. Site selection credit is achieved by not locating the building, or hardscapes, on the list of prohibited sites.

24. *The answer is:* (D) reduced toxic waste

Because mercury waste is toxic, using light bulbs with low mercury content, long life, and high lumen output will result in reduced toxic waste. Mercury content of lights does not affect light pollution, which refers to the impact of artificial light on night sky visibility.

25. *The answers are:* (A) light to night sky
    (B) light trespass
    (D) stormwater mitigation

Site lighting and stormwater mitigation must be addressed when designing a sustainable site. On-site renewable energy can reduce the building's burden on the power grid and is addressed in the Energy and Atmosphere category. Refrigerants affect ozone depletion and are addressed in the Energy and Atmosphere category.

While important, controlling light trespass and light to night sky are not prerequisites for sustainable site design.

26. *The answers are:* (B) evaluate durability risks of project
    (C) incorporate durability strategies into design

The four basic elements of a durability plan are evaluation of durability risks; incorporation of durability strategies into design; implementation of durability strategies into construction; and complete a third party inspection of the implemented durability features.

27. *The answers are:* (A) ensure the technical soundness of the LEED reference guides and training
    (C) resolve issues to maintain consistency across different LEED rating systems

Technical Advisory Groups (TAGs) respond to Credit Interpretation Requests (CIRs) and assist in the development of LEED credits. The Technical Scientific Advisory Committee (TSAC) ensures LEED and its supporting documentation is technically sound while assisting USGBC with complex technical issues.

28. *The answers are:* **(B)** preindustrial material

(C) rapidly renewable material

(D) regionally extracted material

The LEED EBO&M rating system defines the amount of material that must come from post-consumer, pre-industrial, rapidly renewable, or regionally extracted sources in order to earn credit for ongoing consumable purchases. The Rainforest Alliance certifies food, and is not related to ongoing consumables. Salvaged material use and purchase can contribute to earning credit for the sustainable purchases of durable goods and facility alterations, but not for ongoing consumables.

29. *The answer is:* **(A)** LEED CI

LEED for Commercial Interiors (CI) addresses tenant spaces within a building. Both LEED for New Construction & Major Renovation (NC) and LEED for Existing Buildings: Operations & Maintenance (EBO&M) are rating systems that apply to entire buildings. LEED for Core & Shell (CS) addresses buildings that are built with no, or limited, interior buildouts.

30. *The answer is:* **(D)** geographic location

USGBC chapters and regional councils identify which credits are eligible for Regional Priority points based on the needs of each environmental zone. LEED Online determines the region of a project based on its geographic location, which it identifies from the project site's zip code.

31. *The answers are:* **(B)** construction cost

(C) documentation cost

(E) soft cost

Project teams should consider the potential construction, soft, and documentation costs before committing to pursing a particular LEED credit. The application review cost is established regardless of the number or selection of LEED credits pursued, and is based on the building's floor area. Registration cost is the same for every LEED project, varying only depending on if the project is registered by a member or non-member company.

32. *The answer is:* **(C)** LEED EBO&M

Credit Interpretation Requests (CIRs) and appeals are components of every LEED rating system. LEED EBO&M is the only rating system that requires the implementation of performance periods.

33. *The answer is:* **(D)** ANSI/ASHRAE/IESNA 90.1-2007

ANSI/ASHRAE/IESNA 90.1-2007 sets minimum requirements for the energy-efficient design of all buildings except low-rise residential buildings. ANSI/ASHRAE 52.2-1999 addresses air cleaner efficiencies; ASHRAE 55-2004 addresses thermal comfort; and ANSI/ASHRAE 62.1-2007 addresses ventilation.

Practice Exam Part One Solutions

34. *The answers are:* (A) Controllability of Systems
(C) Indoor Chemical and Pollutant Source Control

Project teams intending to increase a building's ventilation will have to consider the implications on the building's commissioning, measurement and verification, and energy performance, all of which will be directly affected.

Controllability of Systems relates more to the thermal comfort and lighting of a building than the building's ventilation. Increasing ventilation will not help or prevent a project team from achieving Indoor Chemical and Pollutant Source Control.

35. *The answer is:* (B) track the movement of wood products from the forest to the building

A chain-of-custody document verifies compliance with Forest Stewardship Council (FSC) guidelines for wood products, which requires documentation of every movement of wood products from the forest to the building.

36. *The answers are:* (D) manufactured within 500 miles of the project site
(E) permanently installed on the project site

For the purposes of the LEED rating system, regional materials are those permanently installed building components that have been extracted, harvested or recovered, and manufactured within 500 miles of the project site.

Regional materials do not need to be recycled or post-consumer materials, nor do they need to be rapidly renewable (agricultural products that take 10 years or less to grow or raise and can be harvested in an ongoing and sustainable fashion). Forest Stewardship Council (FSC) certification applies only to wood and is not a requirement of regional materials.

37. *The answer is:* (C) Materials Reuse

A project team's choice of paint will have little or no effect on materials reuse. Choosing white exterior paint can contribute to reducing a building's heat island effect. Choosing paint with low volatile organic compounds (VOCs) will contribute to earning Low-Emitting Materials credit. Choosing bio-based paint can help a project team earn Rapidly Renewable Materials credit.

38. *The answers are:* (A) core credits
(C) innovation credits
(E) prerequisites

All LEED rating systems contain prerequisites, core credits, and innovation credits. Sustainable operations and educational programs may help a project team achieve either a core credit or an innovation credit.

39. *The answer is:* (B) door

Computers, office desks, and landscaping equipment are examples of durable goods, which are defined by the LEED reference guides as goods with a useful life of two years or more and that are replaced infrequently. Doors are considered part of the base building equipment.

40. *The answers are:* (C) local climate

                           (D) operational costs

Installing larger ventilation systems will minimally impact construction costs, but will significantly increase the energy cost throughout the life cycle of the building. Prior to increasing the ventilation rate of a building, the design winter and summer temperatures and humidity should be considered. Regardless of building location, the interior temperature should typically be between 68 and 74°F and the relative humidity should be between 50 and 55%. Refrigerant management is not affected by ventilation systems.

41. *The answers are:* (B) global warming potential

                           (C) ozone depletion potential

Refrigerants are chemical compounds that, when released to the atmosphere, deteriorate the ozone layer and increase greenhouse gas levels. LEED requires project teams to consider the ozone depletion potential and the global warming potential of refrigerants used in a building's HVAC & R system.

42. *The answers are:* (B) email address

                           (C) industry sector

                         (E) phone number

There is no cost to set up a personal user account on the USGBC website. The individual does not need to be a USGBC member, have prior LEED project experience, have or intend to have LEED credential, or supply his or her company name.

43. *The answers are:* (B) CIRs reviewed by a TAG for their own project

                           (D) CIRs posted prior to project application, for the applicable rating system only

Project teams are required to adhere only to CIRs uploaded to the USGBC website prior to project registration. Adherence is required for those submitted after project registration only if they were submitted by the project team itself. CIR requirements must be adhered to regardless of the project's geographic location; however, project teams are generally only required to follow CIRs for their specific rating system.

44. *The answers are:* (C) achieve the next incremental level of an existing credit

                         (D) double the requirements of an existing credit

Innovation in Design points for exemplary performance are earned for going above and beyond existing credit requirements. Alternatively, project teams can earn ID points for achieving the next incremental level of an existing credit if it is specified within the corresponding rating system.

45. *The answers are:* (B) collecting rainwater for sewage conveyance

                         (D) using cooling condensate for cooling tower make-up

As with every sustainable strategy, consult all applicable codes prior to implementation. Non-potable water used for cooling tower make-up and sewage conveyance can both lead to earning

LEED credit. On-site septic tanks do not reduce potable water used for sewage conveyance, and therefore do not contribute to the achievement of a LEED point for water conservation. Code requirements restrict project teams from using blackwater for landscape irrigation.

46. *The answer is:* **(C)** LEED credit equivalence submittal

Project teams intending to achieve Innovation in Design credit for innovation (not exemplary performance) must follow the LEED credit equivalence process, which requires the following.

- the proposed innovation credit intent
- the proposed credit requirement for compliance
- the proposed submittal to demonstrate compliance
- a summary of potential design approaches that may be used to meet the requirements

The Credit Interpretation Requests (CIRs) process does not involve the proposition of new credits, and therefore does not require an explanation of a proposed credit requirement. The LEED Online submittal templates do not allow for the proposal of new credits.

47. *The answer is:* **(B)** Center for Resource Solutions

The Center for Resource Solutions' Green-e energy program is a voluntary certification and verification program for renewable energy products.

Energy Star's Portfolio Manager is a federal program that helps businesses and individuals protect the environment through energy efficiency. The Department of Energy's mission is to advance the energy security of the United States. There is no such thing as the Center for Research and Development of Green Power.

48. *The answers are:* **(B)** International Energy Conservation Code (IECC)
(D) International Mechanical Code (IMC)
(E) International Plumbing Code (IPC)

The International Code Council (ICC) is a consolidated organization that comprises what was formerly the Building Officials and Code Administrators International, Inc. (BOCA), the International Conference of Building Officials (ICBO), and the Southern Building Code Congress International, Inc. (SBCCI). The ICC family of codes includes, but is not limited to, the International Building Code (IBC), the International Fire Code (IFC), the International Plumbing Code (IPC), the International Mechanical Code (IMC), and the International Energy Conservation Code (IECC).

49. *The answer is:* **(C)** Companies can be USGBC members.

People can be LEED-accredited, buildings can be LEED-certified, and companies (not individuals) can be USGBC members. It is not a requirement that a LEED project administrator (or anyone else involved with a LEED project) be a LEED AP.

50. The answer is: (B) 80%

A project team earning more than 40% but less than 50% of core credits within the LEED rating systems will earn a LEED Certified plaque. They will earn a LEED Silver plaque for earning more than 50% but less than 60% of Core Credits; LEED Gold for earning more than 60% but less than 80%; and LEED Platinum for earning more than 80%.

51. The answers are: (C) a means to submit a project for review
(E) the project's final scorecard

Project teams submit their projects to the Green Building Certification Institute (GBCI) for review using the LEED submittal templates, which also generate the final scorecard for LEED projects.

Project team members can manipulate the documentation for only those prerequisites or credits assigned to them. The LEED reference guides contain potential strategies for credit and prerequisite approval. Credit Interpretation Requests are reviewed and submitted at LEED Online, but not though submittal templates.

52. The answers are: (A) bid
(B) design
(D) pre-design

According to the *Sustainable Building Technical Manual*, an environmentally responsive design process includes the following key steps: pre-design, design, bid, construction, and occupancy. While the selection of vendors, consultants, and/or contractors is part of the process, it is not a key step. Re-bid and post-design should not occur if an appropriate process has been followed.

53. The answer is: (C) Projects can earn half points under the Site Selection credit.

LEED EB and LEED EBO&M projects must recertify at least every five years to maintain their LEED certification, but the LEED CI rating system does not have a recertification option. Precertification is only available if a project team is utilizing the LEED for Core & Shell rating system. Prohibiting smoking is an option in every LEED rating system for meeting the Environmental Tobacco Smoke prerequisite; because this is a prerequisite, no points are awarded for compliance.

54. The answers are: (B) clarify
(D) denied
(E) earned

An *earned* ruling requires no additional action from the project team. A *clarify* ruling requires the project team member to address the issues of the project reviewer. A *denied* ruling indicates that the project team member either misunderstood the intent and/or failed to meet the prerequisite or credit. Anticipated and deferred are ways to categorize prerequisites and credits reviewed during the design phase submittal.

55. *The answers are:* (A) delegate responsibility and oversee all LEED committee activities
(D) establish and enforce LEED direction and policy

The role of the LEED Steering Committee is to establish and enforce LEED direction and policy as well as to delegate responsibility and oversee all LEED committee activities.

Technical Advisory Groups (TAGs) respond to Credit Interpretation Requests and the GBCI develops and administers the accreditation exams. The Technical Scientific Advisory Committee (TSAC) ensures LEED and its supporting documentation is technically sound.

56. *The answers are:* (B) assign credits to project team members
(C) build a project team
(E) submit projects for review

Project administrators can do many things through the LEED Online workspace, including assigning credits to project team members, building the project team, and submitting projects for review. Previously submitted Credit Interpretations Requests (CIRs) can be viewed by USGBC company members as well as LEED project team members; however, credit appeals submitted by other project teams are not available for review. Precertification is a unique aspect of the LEED for Core & Shell rating system.

57. *The answers are:* (E) LEED for Homes rating system
(F) sample LEED submittal templates
(G) usgbc.org account

A free download of the LEED for Homes rating system is available at www.usgbc.org/homes. Sample LEED submittal templates are available at LEED Online. Individuals may create a free usgbc.org account.

All Credit Interpretation Requests (CIRs) require a fee be submitted to USGBC before a Technical Advisory Committee (TAG) will review it. The *LEED Reference Guide for Building Design and Construction* is available for purchase at **www.ppi2pass.com/LEED**. USGBC and LEED brochures are available for purchase at www.gbci.org/publications. LEED certification requires project registration, whose fees are described at www.usgbc.org/leedregistration.

58. *The answers are:* (A) additional response time may be incurred
(C) it is sent to the LEED Steering Committee

CIRs beyond the expertise of the Technical Advisory Group (TAG) are sent to the LEED Steering Committee and/or relevant LEED Committees for a ruling. Additional time is typically incurred in this situation.

59. *The answer is:* (A) LEED project tools

LEED project tools are only available to registered project teams. A directory of all LEED registered and certified projects, as well as posted rulings on Credit Interpretation Requests are available to registered USGBC members. LEED Online does not contain a list of LEED project credit appeals. While the project's submittal templates also become available after the project is registered, sample submittal templates are available to the public.

60. *The answers are:* (A) codes and regulations that address asbestos and water discharge
(B) codes and regulations that address PCBs and water management

Buildings must be in compliance with federal, state, and local environmental laws and regulations, including but not limited to those addressing asbestos, PCBs, water discharge, and water management. LEED certification can be revoked upon knowledge of noncompliance.

Complying with referenced standards that address sustainable forest management practices, fixture performance requirements for water use, and volatile organic compounds (VOCs) may help achieve LEED credits; however, doing so is not a they are not program requirement.

61. *The answers are:* (A) include members from varying industry sectors
(E) involve the team in different project phases

An integrated project team actively involves participants from varying industry sectors throughout the project, and meets monthly to review the project status.

The role of the commissioning authority is to verify the mechanical systems are operating as the designer intended. Product vendors may provide valuable insight that promoted the success of the project; however, they are not required on the project team. The purpose of including a LEED AP on a project team is to encourage LEED design integration and to streamline the application and certification processes, but including a LEED AP in not needed to create an integrated project team.

62. *The answer is:* (C) LEED EBO&M

Facilities undergoing minor envelope, interior, or mechanical changes must pursue LEED certification under the EBO&M rating system. Facilities undergoing major envelope, interior, or mechanical changes must pursue LEED certification under the NC rating system. LEED CI is used to certify tenant spaces while LEED CS is used to certify buildings prior to tenant infill.

63. *The answers are:* (D) use rainwater for sewage conveyance or landscape irrigation
(E) remediate contaminated soil

Soil remediation may help a project team achieve a credit under Brownfield Redevelopment, and using collected rainwater may earn a team points under the Water Efficiency category.

Erosion and sedimentation control, installing HCFC-based HVAC & R equipment, and the storage and collection of recyclables are LEED certification prerequisites (not credits).

64. *The answer is:* (D) To achieve LEED certification, every prerequisite and a minimum number of credits must be achieved.

The first step toward LEED certification is project registration. A fee must be paid when submitting Credit Interpretation Requests (CIRs). There is a LEED rating system for every type of building, including non-commercial buildings.

65. *The answer is:* (B) Optimize Energy Performance

Ground source heat pumps are energy efficient mechanical systems that may help project teams earn the energy performance prerequisite and credit. Ground source heat pumps contain a vapor compression refrigeration cycle, which requires electricity to operate, and therefore it is not a renewable source of energy. They do not generate energy (so they don't contribute to On-Site Renewable Energy credits), nor do they reduce the amount of water used by a project.

66. *The answers are:* (D) offer a volume certification path
    (E) provide a streamlined certification process for large-scale projects

The goals of the Portfolio Program are to assist participants in integrating green building design, construction, and operations into their standard business practices using LEED technical standards and guidance; provide a cost-effective, streamlined certification path for multiple buildings that are nearly identical in design; recognize leaders who are creating market transformation through their commitments and achievements in green building; foster a network of investors, developers, owners, and managers committed to systemically greening their building portfolios; and support participating organizations in fulfilling their sustainability commitments by providing solid performance metrics that can be given to stakeholders.

Submittal templates provide a template of key data for the design team members to compile. The mission of LEED in general is to encourage and accelerate the "global adoption of sustainable green building and development practices through the creation and implementation of universally understood and accepted standards, tools, and performance criteria."

67. *The answers are:* (A) before the design phase
    (B) during the construction phase

The LEED project budget should be addressed before the design phase and throughout the construction phase of the project. LEED project team selection is based on of the created budget.

68. *The answer is:* (C) installing native or adapted plants

Turf grass typically requires sizable amounts of water to sustain it, while native or adapted plants can survive on the amount of rainfall a site naturally receives. Projects with no landscaping are not eligible to earn landscaping credits. Invasive plants, such as weeds, are not considered landscaping by the LEED reference guides.

69. *The answer is:* (B) 4 points

Regionalization is an opportunity for project teams to earn additional points in the Innovation in Design category of each rating system. GBCI determines which six credits are priorities in each region of the US. Project teams in each region can earn up to four additional points for achieving the Regional Priority credits assigned to its region.

70. *The answer is:* (B) Energy Metrics

Energy Metrics credits focus on the building's energy performance and ozone protection.

Energy and Atmosphere is a credit category within the LEED rating systems, and is not a functional characteristic grouping. Materials Out credits are associated with the sustainable solid waste management policy of a building. Site Management credits address sustainable landscape management practices.

71. The answers are: **(B)** interior moisture loads
(D) pests
(E) ultraviolet radiation

The intent of durability planning is to appropriately design and construct high performance buildings that will continue to perform well over time. The principle risks are exterior water, interior moisture, air infiltration, interstitial condensation, heat loss, ultraviolet radiation, pests, and natural disasters.

72. The answer is: **(B)** contact a LEED for Homes provider

The first step for participating in the LEED for Homes program is to contact a LEED for Homes provider and then to register the project with GBCI. Becoming a LEED AP will strengthen your LEED knowledge; however, it is not required to pursue LEED certification in any rating system. Credit Interpretation Requests can be submitted only after a project is registered.

73. The answer is: **(D)** refrigerant leakage

HVAC & R equipment with a relatively short equipment life, a relatively high refrigerant charge, and/or refrigerant leakage will contribute to ozone depletion and global warming.

74. The answer is: **(B)** durability plan

As it is explained in the LEED for Homes rating system, a successful durability plan includes a ranking of durability risks and strategies to minimize the risks, which could include sealing ventilation ducts, installing rodent and corrosion-proof screens, and using air sealing pump covers.

The PE exemption form gives project teams the opportunity to follow a streamlined path to achieve certain prerequisites and credits. A building's landscape management plan focuses on ecology and wildlife outside of the building.

75. The answer is: **(A)** ASHRAE 55-2004

ASHRAE 55-2004 was created to establish acceptable indoor thermal environmental conditions. ANSI/ASHRAE 52.2-1999 addresses air cleaner efficiencies; ANSI/ASHRAE 62.1-2007 addresses ventilation; ANSI/ASHRAE/IESNA 90.1-2007 addresses building efficiency.

76. The answers are: **(A)** email address
(D) organization name
(E) individual's title

The primary contact of a LEED project is not required to be a LEED AP or have previous LEED project experience.

77. *The answers are:* (D) project summary
(E) project team members

The project team and project summary must be confirmed prior to the LEED application process. CIRs and precertification are items available for LEED project administrators; however, they are not required items. The project cost does not need to be confirmed prior to registration.

78. *The answer is:* (A) evapotranspiration

Evapotranspiration is the term used to describe water that is lost through plant transpiration and evaporation from soil.

79. *The answer is:* (D) view a protected electronic version of the LEED reference guide

An online reference guide access code is provided with the purchase of a LEED reference guide. The code provides its owner with electronic access to the reference guide. Project team members can also acquire the access code for 30 days of electronic access to the reference guide from a project team administrator after joining a LEED project.

80. *The answer is:* (C) intent

The structure of LEED prerequisites and credits is considered part of the LEED brand and includes intent, requirements, and potential technologies and strategies. This structure is maintained in each version of the LEED rating system. Economic and environmental benefits, greening opportunities, and submittal requirements are included in the rating systems and/or LEED online.

81. *The answers are:* (B) durable goods
(E) waste stream

The environmental impact of refrigerants and energy consumption are addressed in the Energy and Atmosphere category. Materials and Resources credits address the flow of materials to and from a project site as well as the waste stream generated from a project building.

82. *The answer is:* (D) urban area

Building on a previously developed site helps conserve existing greenfields (undeveloped land). Locating a project in an urban location will provide the occupants with increased opportunities to utilize public transportation and reduce the need to drive to local amenities. LEED prohibits projects from building on public parkland, due to the potential economic and environmental implications.

83. *The answer is:* (A) declarant's name

Every LEED submittal template begins with the following statement: "I, *declarant's name*, from *company name* verify that the information provided below is accurate, to the best of my knowledge."

Product manufacturer info may be required for some documentation; however, it is not a requirement of every submittal. The project location is identified in the LEED registration form.

84. *The answers are:* **(B)** establishing a project budget

    (C) establishing project goals

    (D) site selection

The pre-design phase of every LEED project includes developing a green vision, project goals, priorities, building program, and budget; assembling a green team; developing partner strategies and project schedules; researching and reviewing local codes, laws, standards; and selecting a project site. Commissioning, testing and balancing, and training are steps of the construction phase.

85. *The answer is:* **(A)** CIRs must be submitted as text-based inquiries.

There is no mechanism available to submit attachments during the CIR process. A complete project narrative is not required, and the CIR is limited to 600 words.

86. *The answers are:* **(D)** energy efficiency

    (E) environmental impact

    (F) indoor environment

Green building design and construction should be guided primarily by energy efficiency, environmental impact, resource conservation and recycling, indoor environmental quality, and community considerations. Design and bid costs should be considered; however, green building focuses primarily on life-cycle costs. Construction documents are created once the guideline issues have been addressed.

87. *The answers are:* **(A)** accreditation of industry professionals

    (C) certification of sustainable buildings

GBCI administers the LEED accreditation for industry professionals, and is responsible for the certification of buildings and parts of buildings. It does not certify products. USGBC provides sustainable education programs.

88. *The answers are:* **(A)** appeal

    (D) construction

Credits can be earned during the construction phase of a project. Design credits can be submitted in the design review, but they will only be distinguished as "anticipated" or "denied." Project teams can modify a strategy and resubmit a design submittal credit in the construction phase to change the designation from "denied" to "earned". The construction review must include all design and construction credits that are pursued by the project team. After the construction review, credits are designated as "earned" or "denied." This designation can only be changed by submitting an appeal and paying a fee for each credit appealed. After the credit is appealed, it will be designated "earned" or "denied." The credit cannot be appealed a second time.

Practice Exam Part One Solutions

89. *The answers are:* (D) promotes design efficiency
    (E) reduces design and construction time

Commissioning increases a project's initial cost, but should reduce its life-cycle cost. Commissioning authorities are not responsible for ensuring a project's compliance with a code. Rather, they verify that review design documents to help eliminate changes during construction or contractors from making design decisions. The commissioning agent's check helps prevent consultant re-design, as well as on-the-job engineering from the contractors, and therefore reduces the overall time spent on design and construction.

90. *The answers are:* (A) anthropogenic nitrogen oxide
    (B) carbon dioxide
    (D) sulfur dioxide

Conventional fossil-fuel generated electricity (such as that from coal-fueled power plants and liquid petroleum) results in the release of carbon dioxide into the atmosphere, which contributes to global warming. Coal-fired electric utilities emit both anthropogenic nitrogen oxide (nitrogen oxide that is a result of human activities, and which is a key contributor to smog) and sulfur dioxide (a key contributor to acid rain).

91. *The answer is:* (D) shading hardscapes with vegetation

Heat island effect can be minimized by having high (not low) Solar Reflectance Index (SRI) values; minimizing the area of hardscapes; and shading necessary hardscapes with trees and bushes. The glazing factor is related to daylighting, not heat island effect.

92. *The answers are:* (B) indoor environmental quality criteria
    (C) mechanical system descriptions
    (E) references to applicable codes

Mechanical systems (which include HVAC & R, plumbing, and electrical systems) must be addressed in the Basis of Design (BOD). The BOD must also establish the procedure for the installed mechanical equipment to achieve the required indoor environmental quality criteria. Applicable codes must also be included to help provide guidance to the installing contractors. Process equipment and building materials may be addressed here; however, it is not a requirement.

93. *The answers are:* (B) energy sustainability consultant
    (C) landscape architect

Utility managers, product manufacturers, and code officials may be a good resource for sustainability advice; however, they are not directly involved with the decisions of the project, and therefore are not considered part of the integrated project team.

94. *The answer is:* (A) 50%

Replacement or upgrade to 50% of the building's envelope (walls, floors, and roof) is considered a major renovation. Replacement or upgrade to 50% of the building's interior (non-structural walls, floor coverings, and drop ceilings) is considered major. Replacement or upgrade to 50% of the building's mechanical systems (HVAC & R, lighting, plumbing) is considered major. The percentages are calculated using either the cost of the renovation or the area being renovated.

95. *The answer is:* **(B)** 50%

Owners can occupy up to 50% of the building's leasable space and still be eligible to pursue LEED certification under the Core & Shell rating system. Buildings in which the owner occupies more than 50% of the leasable floor space must pursue LEED certification under the New Construction rating system.

96. *The answers are:* **(A)** Energy and Atmosphere
(F) Operational Effectiveness

Credits and prerequisites can be grouped by the credit categories designated within the LEED rating systems, or by functional characteristics. The Best Management Practices prerequisite falls within in the Energy and Atmosphere LEED credit category, but can also be grouped as an Operational Effectiveness prerequisite by its functional characteristic. Operational Effectiveness credits and prerequisites support best management practices.

97. *The answers are:* **(B)** gross area of the building
(D) project budget
(E) site conditions

The project's primary contact is the only individual required to submit their company's name. The project team creates a list of possible innovation strategies, which is submitted during the project application phase.

98. *The answer is:* **(B)** previously certified under the LEED EBO&M rating system

LEED for Existing Buildings: Operations & Maintenance is the only rating system that has a recertification option available. Precertification does not exist for the LEED for Schools rating system.

99. *The answers are:* **(C)** installing heat recovery systems
(E) zoning mechanical systems

Increasing the ventilation rate and performing a flush out before occupancy improves indoor air quality but increases the amount of energy used. Reducing a building's heat island effect will not affect the building's energy use.

100. *The answer is:* **(D)** VOC content in building materials

A building's indoor environmental quality can be improved by controlling noise pollution, providing as much natural lighting as possible, and providing adequate ventilation (thereby improving the air quality). Volatile organic compounds (VOCs), which damage lung tissue, should be minimized.

Practice Exam Part Two Solutions

1. D
2. C, D
3. A
4. A, D
5. A, C, D
6. C
7. A, D, E
8. B, C
9. C
10. A, B, C
11. C
12. B
13. C, D, E
14. B
15. D, F, G
16. C
17. C
18. B
19. C
20. C, D, E
21. D
22. D
23. D
24. A, C, D
25. A, C, E
26. A, B, D
27. B
28. C
29. B, C
30. A, C
31. C
32. B, C, E
33. B
34. D, E
35. C
36. A
37. B
38. D
39. B
40. A, D
41. B
42. B
43. D
44. D
45. A, C
46. A
47. A
48. D
49. D
50. D
51. C
52. B
53. A
54. C
55. A, C
56. C
57. A, C, E
58. B
59. C
60. C
61. B, C, D
62. A, B
63. D
64. B, D
65. B, C, D
66. D
67. B, D, E
68. C
69. A, C, E
70. C
71. C
72. C
73. C
74. B, C, D
75. B, C, D

LEED O&M Practice Exam

76. A
77. C
78. C
79. B, C, D
80. A
81. C
82. B, C, D
83. A, B, D
84. B, C, E
85. C, D, E
86. C
87. A, D, E
88. A, B, D
89. C, D
90. B
91. B
92. D
93. A, C, E
94. A, C, E
95. C
96. A
97. C
98. C
99. B
100. A, D, E

Practice Exam Part Two Solutions

1. *The answer is:* **(D)** 8 points

Reducing the building occupants' vehicle use by 25% will qualify the team for seven points for SS Credit 4.2, Alternative Commuting Transportation. Including one or more LEED AP on the project team earns one point for IO Credit 2, LEED Accredited Professional. An Energy Star rating of 69 is the minimum requirement of EQ Prerequisite 1, Minimum IAQ Performance, and thus is not eligible for point earning. Since the building does not meet EA Credit 4's 25% minimum for off-site energy, it is not eligible to earn a point for this credit.

2. *The answers are:* **(C)** LEED EBO&M 2009 rating system
(D) LEED EB version 2.0 rating system

Project teams must register under the rating system available at the time of project registration which, in this case, was the LEED EB version 2.0 rating system. LEED project teams have the opportunity, however, to use the newest version of the rating system if it becomes available before the performance periods begin. Because the LEED EBO&M 2009 rating system became available before the team submitted its credit for review, it could opt to change to this rating system. There is no charge to change a project's rating system after registration.

Since this project does not include major renovations, the LEED NC rating system cannot be used. Furthermore, the LEED CI rating system applies to individual tenant spaces or parts of a building (not the entire building), so the project team cannot use the LEED CI rating system either.

3. *The answer is:* **(A)** 3 points

Project teams can earn between three and fifteen points for projects that reduce conventional commuting trips for between 10% and 75% of the full-time equivalent (FTE) occupants. The LEED EBO&M rating system allows the team to choose its combination of strategies to reduce conventional commuting trips. Some alternative commuting options available in SS Credit 4 include providing bicycle racks, bicycling equipment, changing facilities, preferred parking, access to mass transit, and/or alternative-fuel refueling stations. Employers can also offer additional vacation days, cash rewards or pretax options, free or discounted public transportation passes, and/or telecommuting equipment. Note that an occupant given more than one benefit will not count more than once.

4. *The answers are:* **(A)** determine the financial impacts of various LEED rating system credits
(D) provide the owner with proof of return on investment for building upgrades

Documenting the operating costs before and after implementing a sustainable strategy can help the owner determine the financial benefits of the credits. This documentation is not intended to help the owner evaluate employee performance, nor is the building's carbon footprint is directly related to documenting sustainable strategy.

5. *The answers are:* (A) EA Prerequisite 1, Energy Efficiency Best Management Practices
   (C) EA Credit 3.1, Performance Measurement: Building Automation System
   (D) EA Credit 3.2, Performance Measurement: System Level Metering

Installing a building automation system contributes toward meeting the requirements of EA Prerequisite 1, Efficiency Best Management Practices, EA Credit 3.1, Performance Measurement: Building Automation System, and EA Credit 3.2, Performance Measurement: System Level Metering.

To achieve EA Credit 6, Emissions Reduction Reporting, the consultants would have to quantify the emissions reduced and report them in a formal tracking system (i.e., Energy Star, EPA Climate Leaders, etc.). Furthermore, verifying that the HVAC & R systems do not contain CFC-based refrigerants is an environmental issue, not necessarily an energy efficiency issue.

6. *The answer is:* (C) major renovations of existing buildings

The LEED EBO&M rating system is designed to certify the ongoing operations and maintenance of existing buildings. It applies to all existing commercial, institutional, and high-rise residential buildings, both public and private, of all sizes. Under the LEED EBO&M rating system, any residential building four or more stories high are high-rise residential buildings. Thus, private high-rise residential buildings, commercial buildings with no previous certification, and institutional buildings over 50,000 square feet are all eligible for certification under the LEED EBO&M rating system.

Major renovation projects must use the LEED NC rating system, not the LEED EBO&M rating system.

7. *The answers are:* (A) area of the building footprint
   (D) total site area
   (E) vegetated site area

The following must be true for SS Credit 5, Site Disturbance: Protect or Restore Open Habitat.

$$(\text{total site area} - \text{area of the building footprint})(0.25) \geq \text{vegetated site area}$$

In other words, a minimum of 25% of the site area, excluding the area of the building footprint, must be covered with native or adapted vegetation. If the building is located in an urban area with no setback, a vegetated roof may apply and a minimum of 5% of the site (including the building footprint) must be vegetated.

8. *The answers are:* (B) exceeding the requirements of an existing credit by a percentage established by the rating system
   (C) implementing a credit-earning strategy from another LEED rating system

Under the LEED EBO&M rating the project teams can earn points for IO Credit 1 when the next incremental level of an existing credit is achieved. Credits from other LEED rating systems not addressed by the LEED EBO&M rating system can also earn a project team points for IO Credit 1. Project teams can earn points for geographically specific strategies through LEED Regional Priority credits.

While having a USGBC member on the team may provide numerous benefits, it does not qualify the project for points. Including a LEED AP on the project team achieves IO Credit 2, LEED Accredited Professional, not IO Credit 1.

9. *The answer is:* (C) site energy usage

A building's site energy is the amount of heat and electricity consumed by a building, as reflected in utility bills.

A building's greenhouse gas emissions vary depending on the energy source and the building's location, and that quantity does not appear on an energy bill. Source energy is the total amount of fuel required to operate a building. On-site renewable energy does not appear on an energy bill.

10. *The answers are:* (A) EA Prerequisite 2, Minimum Energy Efficiency Performance
(B) EA Credit 1, Optimize Energy Efficiency Performance
(C) EA Credit 6, Emissions Reduction Reporting

The Indoor Environmental Quality prerequisites and credits address the air quality inside a building, whereas the Energy and Atmosphere prerequisites and credits address a building's energy usage.

The online Energy Star Portfolio Manager program was created to compare a building's energy usage to that of other similar buildings. This tool can be used to report emissions reduction amounts after energy efficiency measures are implemented.

11. *The answer is:* (C) individual tenant spaces within an existing building

The LEED EBO&M rating system can apply to an entire building regardless of occupancy, or to multiple buildings if they are located on the same site and are owned by the same entity.

Individual tenant spaces are ineligible for LEED certification under the EBO&M rating system and must follow the LEED CI rating system. Up to 10% of a LEED EBO&M project may be excluded from the certification process if it is managed by another owner. Other exemptions are prohibited.

12. *The answer is:* (B) 15% for an average weather year and a two-year, 24-hour design storm

In addition to designing and installing the rainwater collection system, SS Credit 6, Stormwater Quantity Control requires an annual inspection program to confirm the collection of at least 15% of the rainwater falling on the whole project site for both an average weather year and a two year, 24-hour design storm.

13. *The answers are:* **(C)** plan to notify building occupants no more than 24 hours after emergency pesticide application

(D) schedule routine inspection and monitoring

(E) specify circumstances under which emergency pesticide application is acceptable

EQ Credit 3.6's integrated pest management (IPM) plan is intended to minimize the building occupants' exposure to chemicals. Project teams must plan to notify building occupants at least 72 hours (not 24 hours) before planned pesticide application and 24 hours after emergency pesticide application. Project teams must schedule routine inspection, monitor the IPM plan, and specify circumstances under which emergency application of pesticides can be applied.

There are no requirements regarding the choice of pest removal over pesticide application, although doing so would minimize chemical exposure, and thus could be a useful strategy for this credit.

14. *The answer is:* **(B)** HVAC & R equipment contains less than 0.5 pounds of refrigerant or the replacement system has a payback of more than 10 years

Buildings with HVAC & R equipment containing CFC-based refrigerant and more than a 10-year payback for replacement are exempt from this prerequisite. This is because if the payback to replace existing equipment takes more than 10 years, the replacement system is not economically feasible.

Buildings with small HVAC & R equipment using less than 0.5 pounds of refrigerant, such as refrigerators, water coolers, and so on, are also exempt from this prerequisite.

15. *The answers are:* **(D)** green cleaning

(F) renewable energy

(G) sustainable purchasing

Project teams can earn up to three points for exemplary performance under the Innovations in Operations category. Projects located within the United States are eligible to earn up to four points under the LEED EBO&M rating system for implementing strategies that address geographically specific environmental priorities. Under the Energy and Atmosphere category, project teams can earn up to six points for implementing on- and off-site renewable energy strategies, and up to 18 points for energy efficiency. Six points are available under the Indoor Environmental Quality category for implementing various green cleaning strategies. Six points are available for sustainable purchasing under the Materials and Resources category. Fifteen points are available to project teams implementing alternative commuting transportation strategies.

16. *The answer is:* **(C)** registering the project

Registration is the first step required for project certification. Registration allows the LEED project team online access for submitting and receiving project information.

Credit Interpretation Requests are available after project registration. It is not a requirement, but LEED certification fees are reduced if submitted by a company that is a USGBC member. Including a LEED AP on the project team will earn one point towards LEED certification; however, it is not a required step.

Practice Exam Part Two Solutions

17. The answer is: (C) 50%

50% of the roof area less the skylights, mechanical equipment, and appurtenances must be vegetated to be eligible for point earning under SS Credit 7.2, Heat-Island Reduction: Roof.

18. The answer is: (B) ASHRAE 55-2004

ANSI/ASHRAE 62.1-2007 addresses indoor air quality; ANSI/ASHRAE 52.2-2004 deals with filter efficiencies; and ANSI/ASHRAE/IESNA 90.1-2001 is not referenced by LEED EBO&M, but it addresses a building's overall energy efficiency.

19. The answer is: (C) 25 feet

Designated outdoor smoking areas must be at least 25 feet from fresh air intakes, doors, and operable windows. It is good design practice to evaluate if the 25-foot distance is adequate to prevent environmental tobacco smoke (ETS) from entering the building. Increased separation from designated smoking areas and fresh air intakes, doors, and operable windows may be necessary.

Smoking areas may also be located inside the building provided that the smoking rooms have deck-to-deck partitions (or hard-lid ceilings), continuous exhaust, and that the area meet pressure-test requirements of EQ Prerequisite 2, Environmental Tobacco Smoke (ETS) Control.

20. The answers are:
(C) MR Credit 7, Solid Waste Management: Ongoing Consumables
(D) MR Credit 8, Solid Waste Management: Durable Goods
(E) MR Credit 9, Solid Waste Management: Facility Alterations and Additions

To meet MR Prerequisite 2 the solid waste management policy must address waste generated by the building's alterations and additions, durable goods, and ongoing consumables.

The policy does not mention MR Credit 6's waste stream audit. It is good practice for the policy to include sustainable purchasing plans; however, addressing MR Credit 5, Sustainable Purchasing: Food is not a requirement of the prerequisite.

21. The answer is: (D) project registration

After project registration the project team can begin to collect information and perform calculations to meet the requirements of the prerequisites and credits being pursued.

Appeals can be filed after the application review; however, project information or calculations must have already been performed for an appeal to be filed. Project certification is the final step of the LEED process.

22. The answer is: (D) 160%

The LEED EBO&M rating system allows buildings completed before 1993 to comply with less stringent potable water standards than those built after 1993. If the date of substantial completion was in 1993 or later, the baseline is 120% of the Uniform Plumbing Code (UPC) or International Plumbing Code (IPC). The date of substantial completion is defined as either the completion of building construction or the date of the most recent plumbing renovation.

23. *The answer is:* **(D)** 50%

50% of work stations and multi-occupant rooms must have lighting control for project teams to earn a point for controllability of lighting systems.

24. *The answers are:* **(A)** appliances
(C) office equipment
(D) televisions

MR Credit 8, Solid Waste Management: Durable Goods addresses office equipment, appliances, televisions, external power adapters, audiovisual equipment and other miscellaneous durable goods. Even though they may be recycled, toner cartridges do not qualify as a durable good under this credit. Mercury containing light bulbs are addressed in MR Credit 4, Sustainable Purchasing: Reduced Mercury in Lamps.

25. *The answers are:* **(A)** batteries
(C) furniture
(E) lamps

Items that contribute to meeting the requirements of MR Credits 1–5, Sustainable Purchasing, include durable goods, food, lamps, future building renovations or additions, and so on. It may be a wise environmental choice to purchase carbon credits or renewable power; however, purchasing them cannot be use toward a Sustainable Purchasing credit. From a building owner's perspective, purchasing graywater would present potential risks and liability issues.

26. *The answers are:* **(A)** hand hygiene strategies
(B) maintenance staff training
(D) standard operating procedures

A green policy must cover standard operating procedures, hand hygiene strategies, the handling and storage of cleaning chemicals, maintenance staff training, and the occupant survey.

27. *The answer is:* **(B)** call the CIR hotline for assistance

USGBC will not answer calls regarding technical questions. Rather, the credit interpretation request process provides a means of addressing these questions. A typical request will ask the USGBC if a project's planned strategy will achieve the LEED prerequisite or credit.

28. *The answer is:* **(C)** 4 points

One, two, three, four, and five points will be awarded respectively for a 10%, 15%, 20%, 25%, or 30% reduction in potable water usage resulting from the efficiency of the building's plumbing fixtures. Only potable water from wells or municipal water systems can count toward this credit.

Practice Exam Part Two Solutions

29. *The answers are:* (A) baseline of waste amounts
    (C) every ongoing consumable
    (D) opportunities for increased recycling and waste diversion

Durable goods and construction waste generated from alterations do not need to be included in the waste stream audit.

30. *The answers are:* (A) EQ Prerequisite 1, Minimum IAQ Performance
    (C) EQ Credit 1.2, Outdoor Air Delivery Monitoring

EQ Prerequisite 1, Minimum IAQ Performance, requires that the building follow the ANSI/ASHRAE 62.1-2007 ventilation requirements, or if the standard is infeasible due to physical constraints, the system may provide 10 cubic feet per minute of outside air per occupant. EQ Credit 1.2, Outdoor Air Delivery Monitoring, addresses the ventilation system monitoring requirements.

The requirements of EQ Credit 1.1, IAQ Management Program, include ongoing evaluation and correction of IAQ issues and are much more extensive than simply monitoring outside air. EQ Credit 1.3, Increased Ventilation, requires a building's ventilation rates to exceed those set by ANSI/ASHRAE 62.1-2007 by 30%; just meeting these requirements will not qualify the team for this credit.

31. *The answer is:* (C) LEED EBO&M initial certification

The LEED EB and LEED EBO&M rating systems are the only rating systems under which a project can be recertified. Any certification application for a building that already has LEED EB certification is considered recertification. An owner who pursued LEED certification under the NC, CS, or CI rating system and earned LEED Gold certification can subsequently get a Platinum plaque through the LEED EBO&M rating system only. The first time a project pursues LEED EBO&M, regardless of previous LEED certifications under other rating systems, it is considered an initial certification.

32. *The answers are:* (B) design indoor conditions
    (C) mode of operation
    (E) time-of-day schedules

The building operation plan summarizes the intended operation of each base building system, and as such should include the mode of operation, the design indoor conditions, and the time-of-day schedules. The owner's project requirements document the owner's vision of the facility. The basis of design describes how the designers intend to achieve the owner's vision.

33. *The answer is:* (B) 5%, 5 points

The building site area includes the total area within the LEED project boundary. At least 5% of the building site area must be vegetated for a project to receive points for WE Credit 3, Water Efficient Landscaping. If potable water used for landscaping is reduced by 50%, 62.5%, 75%, 87.5%, or 100%, one, two, three, four, and five points will be awarded, respectively.

34. *The answers are:* **(D)** photovoltaic panels

(E) windmills

Geo-exchange type systems such as ground source heat pumps and architectural shading features can help reduce a building's energy usage, but they are not considered renewable sources of energy under the LEED EBO&M rating system. On-site renewable energy systems must be metered appropriately to be eligible for LEED credit.

35. *The answer is:* **(C)** 3 points

Installing native vegetation on a roof helps achieve the following credits, worth one point each.

SS Credit 5, Site Disturbance: Protect or Restore Open Habitat

SS Credit 6, Stormwater Quantity Control

SS Credit 7.2, Heat Island Reduction: Roof

36. *The answer is:* **(A)** building energy consumption costs

The building operating costs during the performance period must be compared to the building operating costs of the previous five years. Project teams must also implement an ongoing tracking program to record building operating costs after the performance period to achieve IO Credit 3, Documenting Sustainable Building Cost Impacts.

37. *The answer is:* **(B)** 75%, 12 months, 10%

For the 12-month period prior to submitting the certification application, the building may not be more than 25% vacant, based on a time-weighted average, except during the two-month precertification application period. For transient type buildings such as hotels, schools, dormitories, and so on, ordinary partial occupancy is permitted. Up to 10% of the floor area can be excluded from the project if its operations are under separate management.

38. *The answer is:* **(D)** 6 points

If 3%, 4.5%, 6%, 7.5%, 9%, or 12% of the building's total energy usage is provided from an on-site renewable source, one, two, three, four, five, or six points will be awarded respectively for EA Credit 4, Renewable Energy.

39. *The answer is:* **(B)** EA Credit 5, Enhanced Refrigerant Management

The Montreal Protocol is an international treaty requiring the complete phase-out of CFC and HCFC refrigerants by the year 2030. Because EA Prerequisite 3, Fundamental Refrigerant Management, requires that no CFC-based refrigerants be used in base building HVAC & R systems, and EA Credit 5, Enhanced Refrigerant Management, requires that HCFC-based refrigerants be minimized, compliance with the Montreal Protocol would result in compliance with both.

EA Prerequisite 1, Energy Efficiency Best Management Practices, and EA Credit 6, Emissions Reduction Reporting, do not address refrigerant type requirements.

Practice Exam Part Two Solutions

40. *The answers are:* (A) facility manager
    (D) property manager

The declarant is the member of the LEED project team authorized to prepare and upload the required documentation for a given prerequisite or credit. Not all prerequisites and credits have specific requirements regarding who the declarant must be, but for EQ Credit 3.2, the facility manager or the property manager must be declarant.

41. *The answer is:* (B) development footprint

The development footprint includes hardscapes, access roads, parking lots, non-building facilities, and the building footprint within the project boundary. The property boundary is the entire site area that is under the same ownership as the building. The project boundary is the site area that the project team includes within their LEED certification documentation. The building footprint is the site area that is occupied by the building.

42. *The answer is:* (B) minimize light pollution

Installing low-reflectance surfaces with low SRI values such as black asphalt increases a project's heat island effect while reducing the amount of light sent to the night sky.

43. *The answer is:* (D) as often as every year and at least every 5 years

Project teams must apply for recertification under the version of the rating system current at the time of recertification. Projects are eligible for recertification every year, and recertification must happen at least once every five years. USGBC will not visit the certified building to physically remove the certification plaque if the owner opts not to recertify the building after 5 years; however, they will remove the project name from the USGBC website.

44. *The answer is:* (D) 90%; 90 picograms

90% of all indoor, outdoor, hard-wired, and portable fixtures used for the building or the building's site must be included in the purchasing program. The LEED EBO&M rating system addresses lights that are purchased during, but not prior to, the performance period.

45. *The answers are:* (A) EA Credit 1, Optimize Energy Efficiency Performance
    (C) EA Credit 3.1, Performance Measurement: Building Automation System

Installing a building automation system is one strategy building owners can use to improve energy performance and achieve EA Credits 1 and 3.1. EA Credit 2, Existing Building Commissioning, and EA Credit 5, Enhanced Refrigerant Management, are related to building automation systems, but their implementation is not directly affected by it. EQ Credit 5, Controllability of Systems: Lighting, deals with occupant control, not automated control.

46. *The answer is:* (A) chain-of-custody

Chain-of-custody is a tracking procedure for documenting the status of a product from the point of harvest or extraction to the ultimate end use, including all stages of processing, transformation, manufacturing, and distribution.

Churn is the relocation of people and their workspaces within a space. *Energy audits* identify the energy usage of a building. *Fairtrade* is a certification system that identifies products as having met environmental, labor, and development standards.

47. *The answer is:* **(A)** the time during which operations performance is measured

Before the performance period can begin, the LEED project team must design for and create sustainability policies to comply with all LEED prerequisite and pursued credits. The designs and policies are then implemented and the performance periods can begin. The project team may not stop measuring performance for longer than seven days during the performance periods.

48. *The answer is:* **(D)** 25%

25% of the total food and beverages cost during the performance period must be produced within 100 miles of the building pursuing LEED certification, if this is the only sustainable food purchasing strategy the team implements. The food could also have one of various sustainability certifications to earn the LEED credit.

49. *The answer is:* **(D)** 4 points

The LEED EBO&M rating system awards owners and developers four points when projects are housed in either of two building types: buildings certified under the LEED NC or LEED CS rating systems with at least 75% of the buildout LEED CI compliant.

50. *The answer is:* **(D)** property manager

The property manager oversees operations, maintenance, and upkeep of the building and serves as a liaison between the owner and the tenants, whereas the facility manager is directly responsible for ensuring the functionality of the built environment only.

The building engineer is a qualified engineering professional responsible for the operation and maintenance of the building's mechanical systems.

A company's general manager is responsible for managing both the revenue and cost elements of a company's income statement.

51. *The answer is:* **(C)** utility bill analysis

An energy audit identifies the amount of energy used in a building for what purposes, and identifies opportunities for improving efficiency and reducing costs. ASHRAE uses three levels of energy audits: walk through analyses, energy survey and analysis, and investment-grade audit. An investment-grade audit is a detailed analysis of capital-intensive modifications.

52. *The answer is:* **(B)** 24 months

Along with the summary of the overall ongoing commissioning plan, a schedule verifying that the cycle will be no longer than 24 months must be uploaded to the USGBC website.

Practice Exam Part Two Solutions

53. The answer is: (A) 7 days; 60 days

Consistent start times and durations of performance periods are preferred but not required. The LEED EB version 2.0 rating system required the performance periods to end within three months of each other, but the requirements are more stringent under the LEED EBO&M 2009 rating system.

54. The answer is: (C) 75%

To qualify for MR Credit 8, Solid Waste Management: Durable Goods, the 75% diverted durable goods must have fully depreciated, reached the end of their useful lives for normal business operations, and been removed from the building, site, or organization.

55. The answers are: (A) cleaning agents for site hardscapes
(C) lawn mowers

The building exterior and hardscape management plan must address maintenance equipment, snow and ice removal, cleaning of building exterior, paints and sealants used on building exterior, and cleaning of site hardscapes.

The building's interior items and recycling locations do not need to be included in the plan. Pavement coatings and roof drainage systems are addressed in SS Credit 7, Heat Island Reduction.

56. The answer is: (C) a principle participant on the project team

The LEED AP must be a principle member of the project team; however, the LEED AP is not required to be accredited under the same rating system or version as the project, to be a member of the USGBC, or to submit any specific number of LEED submittal templates.

57. The answers are: (A) building and site
(C) building occupancy and use
(E) project summary and scope

The following must be described in the project narrative: the project summary, scope, and location; the building context, footprint, site area, and parking area; the percent of occupied space; and the occupancies of each space. The project summary and scope should describe the project team's motivation to pursue LEED certification.

The building history and project challenges can be provided, but are not required in the overall project narrative.

58. The answer is: (B) carpet care products

The LEED O & M reference guide provides a list of sustainable criteria that at least 30% of janitorial cleaning products and material must meet to earn a point for EQ Credit 3.3, Green Cleaning: Purchase of Sustainable Cleaning Products and Materials. Included in that list are carpet care products.

Cleaning equipment is included under EQ Credit 3.4, Green Cleaning: Sustainable Cleaning Equipment, not EQ Credit 3.3. The purchase of air cleaning devices and sealants are covered by other credits within the LEED EBO&M rating system.

59. The answer is: (C) 30%

ANSI/ASHRAE 62.1-2007 specifies the minimum ventilation rates and indoor air quality to ensure comfort of the building's occupants. To earn a point for EQ Credit 1.3, Increased Ventilation, the project must provide 30% more outside air than is established by the standard.

60. The answer is: (C) Energy Star Portfolio Manager

The Energy Star Portfolio Manager is a tool that helps project teams compare a building's energy consumption to that of similar buildings. This online tool can also be used to establish energy saving goals and track the savings.

61. The answers are: (B) enhance natural diversity and protect wildlife
(C) integrate high performance building operations into the surrounding landscape
(D) preserve ecological integrity

Eliminating light trespass from the site is the intent of SS Credit 8, Light Pollution Reduction. Reducing heat islands from hardscapes is the intent of SS Credit 7.1, Heat Island Reduction: Non-Roof.

62. The answers are: (A) the number of off-site renewable energy certificates to be purchased
(B) the number of on-site renewable energy certificates to be generated

The Energy Star Portfolio Manager tool evaluates a building's energy use. The percentage of on-site and off-site renewable energy generated or purchased to achieve the credit does not vary. The actual number of certificates depends on the building's energy use. The tool will determine the number, not the percentage, of certificates needed.

63. The answer is: (D) washroom facilities

The LEED O & M reference guide uses the term *usage group* to refer to the population using a certain set of plumbing fixtures. If all fixtures have the same flow rate, or every occupant has access to every washroom within a building, then there is only one usage group.

64. The answers are: (B) air temperature
(D) humidity

Air temperature and humidity must be continuously monitored for EQ Credit 2.3, Occupant Comfort: Thermal Comfort Monitoring, while air speed and radiant temperature can be periodically tested to achieve the credit.

Practice Exam Part Two Solutions

65. *The answers are:* (B) when alterations will affect more than 50% of the building's floor area

(C) when more than 50% of the building's occupants will relocate due to alterations

(D) when the total building area will be expanded by more than 50%

A permanent change in occupancy should not affect a project team's decision to switch from LEED EBO&M to LEED NC. Temporary changes, such as relocation of 50% of the building's occupancy, however, should affect that decision. Furthermore, the decision to switch from LEED EBO&M to LEED NC can be influenced by the extent of changes to the building itself, but not to the project site as a whole.

66. *The answer is:* (D) MERV 13 filters filtering 100% of the outside air and 100% of the return air

MERV 13 filters are capable of capturing smaller particulate matter than MERV 8 filters, so the LEED EBO&M rating system requires that project teams use MERV 13 filters. The installed MERV 13 filters can earn the team a point for EQ Credit 1.4 only if they are filtering 100% of both the outside air and the return air.

67. *The answers are:* (B) cooling tower make-up

(E) landscaping

(F) showers and faucets

Reductions in potable water use for cooling tower make-up, showers and faucets, and landscaping can help achieve Water Efficiency credits. Since it would be unfair to regulate the amount of water people drink, potable water used for drinking fountains is not addressed by any of the Water Efficiency credits. It is good design practice to evaluate the water usage of dishwashers and clothes washers; however, this is not addressed by these credits either. Project teams may pursue a point in the Innovation in Operations category if there are significant savings of potable water used for dishwashers and clothes washers, but these savings will not result in point earning for the WE credit.

68. *The answer is:* (C) 500 gallons

The default ratio of women to men is 50%. On average, women use the water closet three times a day, men use it once a day, and men use the urinal twice a day. The calculation is as follows.

(100 women)(3 uses of the water closet per day)
$\times$ (1 gallon of water per flush) = 300 gallons of water per day

(100 men)(1 use of the water closet per day)
$\times$ (1 gallon of water per flush) = 100 gallons of water per day

(100 men)(2 uses of the urinal per day)
$\times$ (0.5 gallons of water per flush) = 100 gallons of water per day

500 gallons of water per day

69. *The answers are:* (A) building engineer
(C) facility manager
(E) property manager

Some LEED EBO&M prerequisites and credits identify who on the project team may be a *declarant*, or the person responsible for submitting LEED documentation. The declarant is usually the person in charge of implementing or overseeing a particular LEED strategy, and has the technical qualifications to verify the submittal.

The EBO&M rating system specifies that WE Credit 4 may be submitted by the building engineer, facility manager, or property manager. The groundskeeper is not directly involved with cooling tower water management, design, installation, or maintenance, and therefore cannot submit credit documentation. The building owner is not identified as a declarant for any LEED EBO&M prerequisites or credits.

70. *The answer is:* (C) shield all exterior light fixtures 50 watts and over

To earn SS Credit 8, Light Pollution Reduction, either all lights 50 watts and over must be shielded or exterior lighting must not increase perimeter illumination levels by more than 20% when turned on.

Solar powered lights will increase the building's energy performance; however, they may still pollute the night sky, and therefore they do not necessarily qualify for this credit. If 100-watt lights are automatically turned off, any remaining lights will still be on and unshielded, so this strategy alone will not earn the project team any points for this credit.

71. *The answer is:* (C) ongoing consumables and reduced mercury in lamps credits only

MR Prerequisite 1, Sustainable Purchasing Policy, requires that the product purchasing policy address MR Credit 1, Sustainable Purchasing: Ongoing Consumables, along with at least one of the following credits.

MR Credit 2, Sustainable Purchasing: Durable Goods

MR Credit 3, Sustainable Purchasing: Facility Alterations and Additions

MR Credit 4, Sustainable Purchasing: Reduced Mercury in Lamps

MR Credit 5, Sustainable Purchasing: Food, does not need to be addressed in the purchasing policy.

72. *The answer is:* (C) 4 points

The LEED EBO&M rating system limits point earning to a maximum of four points for Innovation in Operations credits. Up to three points can be earned for IO credits achieved by exceeding the requirements of an existing credit by an amount defined by the rating system, meeting the requirements of more than one compliance path of a credit, or meeting the requirements of a credit from a different LEED rating system. Up to four points can be achieved for strategies not defined by the LEED rating systems.

73. The answer is: (C) turf grass on site equal to at least 25% of the project site area

SS Credit 5, Site Development, can be achieved by providing 25% of the site with native or adapted vegetation. If the building has no setback, the project team can provide roof vegetation (native or adapted) equal to 5% of the LEED project site to achieve the credit, or a project team can install native vegetation off site equal to 50% of the site area. Monoculture plantings (i.e., turf) cannot contribute to credit requirements.

74. The answers are: (B) is 70% salvaged
(C) is made of 50% rapidly renewable material
(D) is made of at least 50% FSC-certified wood

The purchase of furniture during the performance period can contribute toward credit earning if one of the following is true of 40% of the total durable good purchase of furniture.

- It is 10% post-consumer or 20% post-industrial.
- It is made of 70% salvaged material off site or on site.
- Its material is 50% rapidly renewable.
- Its wood is 50% FSC-certified.
- Its material is 50% harvested, extracted, and processed within 500 miles of the project.

75. The answers are: (B) minimum narrative requirements
(C) offline credit calculators
(D) policy, program, and plan models

Offline credit calculators supplement the submittal templates for EA Credit 1, EQ Prerequisite 1, and IO Credit 3, and are available on the LEED Registered Project Tools page. The policy, program, and plan models describe the LEED EBO&M policies and provide guidance for their successful implementation. The minimum narrative requirements resource makes recommendations to help project teams submit narratives that will be approved.

The member-to-member exchange is available to every USGBC member, not just registered project teams. The USGBC website contains a list of previously approved exemplary performance credits. This list is available to the public.

76. The answer is: (A) controlled modifications in the lighting levels during the day hours

Daylighting helps provide a stimulating and productive environment while reducing electricity required for the facility's lights. Increased glazing areas, however, can negatively impact the building envelope.

77. The answer is: (C) 50%

At least 50% of the cooling tower's make-up water must come from a non-potable source for WE Credit 4.2, Cooling Tower Water Management: Non-Potable Water Source Use. Additionally, a measurement program must verify the amount of non-potable water available for cooling tower use.

If 100% of the make-up water comes from a non-potable source, an additional point may be achieved for IO Credit 1.

78. The answer is: (C) EA Credit 1, Optimize Energy Efficiency Performance

Installing energy recovery ventilators is a great way to minimize the building's energy use while increasing the outside air brought into the building.

Recovery ventilators do not contain refrigerants, nor do they directly relate to previous LEED certifications of the building or to renewable energy.

79. The answers are: (B) HVAC & R protection
(C) pathway interruption
(D) source control

SMACNA's *IAQ Guidelines for Occupied Buildings Under Construction* recommends that control measures be implemented to address five areas including HVAC & R protection, source control, pathway interruption, housekeeping, and scheduling.

Addressing the VOC limits and cleaning agents is good environmental practice, and may help achieve LEED credits, but *IAQ Guidelines for Occupied Buildings Under Construction* does not require it.

80. The answers are: (A) SS Credit 1, LEED Certified Design and Construction
(C) EA Credit 5, Enhanced Refrigerant Management

The streamlined path exists so that prerequisites and credits previously approved under another LEED rating system will be reapproved with minimal effort on the part of the project team. The appropriate LEED project team member must verify continued compliance and sign the appropriate submittal template.

SS Credit 3 and MR Credits 5 and 7 are unique to the LEED EBO&M rating system and are thus are not available for the streamlined path. The project team may, however, pursue Innovation in Operations points for these credits.

81. The answer is: (C) 30% of the full-time occupants

At least 30% of the full-time occupants must complete the survey. The respondents must include both female and male occupants, and be located on each building exposure (north, south, east, west), in both interior and exterior zones, and in different occupancies (office, reception, shipping and receiving, and so on). Transient occupants can be surveyed; however, it is not required.

82. *The answers are:* (B) investigation and analysis
 (C) implementation
 (D) ongoing commissioning

The first step toward successful commissioning is proper documentation of the current building operations plan. However, this is not a credit requirement, but a prerequisite. Equipment replacement may be part of a building's energy optimization plan but it is not a requirement of the LEED credits associated with commissioning.

83. *The answers are:* (A) calculate the landscaped area of site
 (B) estimate the annual rainfall of site location
 (D) estimate the available graywater for irrigation

Metering potable water used for irrigation is good design practice; however, it is not necessary for WE Credit 3, Water Efficient Landscaping. Blackwater, or water from toilets and urinals, cannot be used for irrigation unless it is cleaned to tertiary standards prior to irrigation.

Graywater can be used to irrigate landscaping in most municipalities, and being aware of its availability can be useful for this credit. Annual site rainfall estimations will help determine how much additional water will be needed to water the landscaping. The landscape architect will need to know the landscaped area, because if it is less than 5% of the site's area the team will not be eligible to earn this credit.

84. *The answers are:* (B) facility manager
 (C) groundskeeper
 (E) property manager

The civil engineer designs the site while the landscaper maintains the landscaping. The declarant of SS Credit 2, Building Exterior and Hardscape Management Plan, must be one of the individuals responsible for site maintenance.

85. *The answers are:* (C) registered architect
 (D) registered landscape architect
 (E) registered professional engineer

Registered professional engineers, architects, and landscape architects have the authority to submit minimal documentation on select credits under the LEED EBO&M rating system. The professional must be part of the LEED project team and the Licensed Professional Exemption Form must be properly completed to pursue the streamlined path.

Neither LEED project team administrators nor commissioning authorities can complete the exemption form unless they are registered professional engineers, architects, or landscape architects.

86. *The answer is:* (C) eight of the air handlers

At least 80% of the building's total fresh air intake (or eight of the building's ten air handlers) must be monitored to achieve EQ Credit 1.2, Outdoor Air Delivery Monitoring.

Because intake serving unoccupied spaces such as stairwells is not considered in the calculation of total fresh air intake, it is not required for all air handlers to have an intake monitoring system.

87. *The answers are:* **(A)** aerators in faucets
(D) dual flush water closets
(E) low-flow showers

Installing irrigation meters is a potential strategy for WE Credit 3, Water Efficient Landscaping. Drift eliminators and conductivity meters that adjust bleed rate both reduce water loss and conserve resources, but they apply to WE Credit 4, Cooling Tower Water Management.

Installing dual-flush water closets, aerators in faucets, waterless urinals, low-flow showers, low-flow faucets, and faucet meters will contribute to WE Credit 2, Additional Indoor Plumbing Fixture and Fitting Efficiency.

88. *The answers are:* **(A)** customers
(B) full-time employees
(D) part-time employees

The full-time equivalent (FTE) calculation includes both full- and part-time occupants as well as transient occupants (visitors, students, customers, and so on). Students, however, are only included in the FTE calculation if the building being certified under the LEED EBO&M rating system is a school or if classrooms are installed. Because landscape maintenance crews do not enter the building, they are not included in the calculation.

89. *The answers are:* **(C)** CIRs are collected twice a month
(D) the TAG contacts the LEED project team administrator with the ruling

Rulings on Credit Interpretation Requests (CIRs) from previous rating systems or for previous versions of the project's rating system do not necessarily apply to other rating systems and versions, so the project team reviews rulings from the LEED rating system and version applicable to their project. If no comparable situation or resolution is found in the CIR database, the project team must submit its CIR as early as possible in the process, and certainly before credit submission with the certification application.

A Technical Advisory Group (TAG) has been created for each LEED credit category. The TAG members are professionals throughout the country who collectively discuss CIRs and respond with a ruling. The TAG does not contact the LEED project team administrator directly. Rather, the TAG's ruling is sent to the project team via e-mail and is posted to the USGBC website within 12 business days of receipt from the TAG.

90. *The answer is:* **(B)** 69

If 50% or more of a building is considered to be a warehouse, the project team must use the Energy Star Portfolio Manager tool for EA Prerequisite 2, Minimum Energy Efficiency Performance. The national average Energy Star rating is 50; a minimum rating of 69 must be achieved to achieve EA Prerequisite 2 and LEED certification; a rating of 71 or higher qualifies the building for LEED point earning; a rating of 75 qualifies the building to be considered an

Energy Star building; and a rating of 95 or higher qualifies the building to earn EA Credit 3's maximum of 18 points.

91. The answer is: (B) 50%, minimum, every other year

Solar Reflectance Index (SRI) is the ability of a surface material to reflect sunlight. The higher the SRI the more solar heat the surface reflects (rather than absorbs). To contribute to this credit, paving materials and parking lot covers have an SRI value of at least 29. Although it is good design practice to clean the reflective surfaces yearly, the requirement for SS Credit 7.1, Heat Island Reduction: Non-Roof, is to clean the surfaces every other year.

92. The answer is: (C) tons of carbon dioxide

Achievement of EA Credit 6, Emissions Reduction Reporting requires the project team to provide the total annual greenhouse gas emissions generated in tons of carbon dioxide for a baseline year and during the performance period.

93. The answers are: (A) Food Alliance
(C) Marine Stewardship Council
(E) Protected Harvest

Food with Food Alliance certification, the Marine Stewardship Council Blue Eco-Label, and Protected Harvest certification, as well as Fair Trade and USDA Certified Organic labels are referenced by MR Credit 5, Sustainable Purchasing: Food. Food and Drug Administration and National Organic Coalition labels are not referenced in this credit.

94. The answers are: (A) achieving Platinum level LEED certification
(C) having a USGBC company member register the project
(E) the building's floor area

If the project is registered by a USGBC company member, the team will receive a discount on both registration and certification application fees. Certification fees are based on the building's floor area (not the project's site area), and they are refunded if Platinum certification is achieved.

Including a LEED AP on the project team does earn the project a point, but the fees are not affected. Fees are the same regardless of site location, and since there is no alternative to online registration, this does not qualify the project for a discount.

95. The answer is: (C) 10 feet

Mats at least 10 feet long must be in place immediately inside all public entryways to earn a point for EQ Credit 3.5, Indoor Chemical and Pollutant Source Control. If the vestibule is five feet long in the direction of travel, five feet of additional matting must also be installed inside the building. If the vestibule is fifteen feet long in the direction of travel, a minimum of 10 feet must be covered with a mat. It is good design practice to include a spigot and plug at each public building entrance for regular maintenance and cleaning of the mats.

96. The answer is: (A) all of the cooling tower make-up water

WE Credit 1.1, Water Performance Measurement: Whole Building Metering, must be achieved for WE Credit 1.2, Water Performance Measurement: Submetering, to be achieved. This means permanently installing a meter that measures the total water use of the entire building and associated grounds. To achieve WE Credit 1.2, at least 80% (not 50%) of the indoor plumbing fixtures and fittings must be submetered for potable water usage. At least 80% of the potable water used for landscaping, domestic hot water, process water or 100% of the cooling tower make-up water can be monitored in lieu of metering the plumbing fixtures and fittings to achieve this credit. Although metering graywater and/or rainwater is possible, it is not a requirement of this credit.

97. The answer is: (C) regularly occupied spaces, 10-foot

The owner, architect, or lighting designer will need to document the floor area of regularly occupied spaces and calculate the daylight illumination levels taken on a 10-foot grid to achieve EQ Credit 2.4, Occupant Comfort: Daylight and Views.

98. The answer is: (C) 50 points

There are four levels of LEED project certification. The required points for each level are as follows.

Certified	40 – 49 points
Silver	50 – 59 points
Gold	60 – 79 points
Platinum	80 – 110 points

99. The answer is: (B) 10 cu ft per minute

Physical constraints may limit the outdoor air rates of existing ventilation systems. The project team must first verify the systems are functioning properly and then modify to provide at least 10 cubic feet per minute of outside air per occupant.

100. The answers are: (A) building engineer
(D) facility manager
(E) property manager

Under the LEED EBO&M rating system, a building employee must declare that they will oversee and continually maintain the chemical management of the cooling tower water. Since the commissioning authority and the mechanical contractor are not building employees, they cannot be the declarant for WE Credit 4.1, Cooling Tower Water Management: Chemical Management.